CONTEMPORARY BRAIN RESEARCH IN CHINA

CONTEMPORARY BRAIN RESEARCH IN CHINA

Selected Recent Papers on Electrophysiological Topics

Translated from the Chinese and edited by

John S. Barlow, M. D.

Neurophysiologist, Neurology Service
Massachusetts General Hospital

Principal Research Associate in Neurology (Neurophysiology)
Harvard Medical School

and

Research Affiliate in Electrical Engineering
Massachusetts Institute of Technology

CONSULTANTS BUREAU · NEW YORK-LONDON · 1971

Library of Congress Catalog Card Number 78-136986

SBN 306-10844-5

A Division of Plenum Publishing Corporation
227 West 17th Street, New York, N.Y. 10011

United Kingdom edition published by Consultants Bureau, London
A Division of Plenum Publishing Company, Ltd.
Donington House, 30 Norfolk Street, London, W.C. 2, England

Printed in the United States of America

INTRODUCTION

The great majority of papers on brain research that have been published in Mainland China in recent years have appeared in the Chinese language (only a small fraction of the work has appeared in English in the journal Scientia Sinica), and hence they have remained inaccessible to other workers, since there have been no translation programs of publications in this field in Chinese of the types that have existed, for example, for Russian-language materials. Accordingly, most investigators are not aware of the work of their Chinese colleagues in this field. Yet the field has been an active if small one in China, and has covered a variety of topics that include electrophysiology, neurochemistry, neuropharmacology, neuropsychology, and instrumentation. Standard techniques and instruments, a number of Chinese manufacture, have been employed. Moreover, Chinese workers have been quite familiar with the publications of other investigators, as is readily apparent from the bibliographies of the papers (see Index).

It was with the thought that much of the work of our Chinese colleagues in brain research had too long remained inaccessible that I began, in 1966, to translate the papers which eventuated in the present collection. The papers, which concern electrophysiological studies both in animals and in man and to a degree reflect my own interests, were selected from Acta Physiologica Sinica (primarily) and Acta Psychologica Sinica for the five-year period 1962-1966, the latter year being the most recent one for which copies of these journals were available. A list of papers on brain research published in the former journal, for this period, with an index to the institutions from which the papers originated, is included in Appendix A. (A list of the papers on brain research which appeared in English, or occasionally in Russian, in Scientia Sinica from the beginning of publication of that journal in 1952 until 1966 (the last year for which copies were available) is given in Appendix B. I have translated the Russian titles, as well as the ones cited in the papers in the present collection, into English, for the convenience of readers not familiar with the Russian language.)

Since this is perhaps the first undertaking of its kind by a western scientist in the field of brain research for whom Chinese is a learned, and not a native tongue, some comments on the procedures followed for preparing the translations may be in order, with the thought that such comments may be useful to other non-Chinese who may also be inclined to take up the not easy task of translating scientific Chinese in this or other fields. Since the written Chinese language is based on ideograms or characters rather than on an alphabet, special problems are encountered both with respect to pronunciation and with respect to dictionaries. A brief summary of several systems of romanization for indicating the pronunciation of Chinese characters is included in Appendix C; using one of these (the National Romanization System, or Gwoyeu Romatzyh), the full text of each paper was typed out, with the meanings of unfamiliar characters or character-groups added above the respective romanized form, the romanization and/or meanings being established as necessary with the aid of dictionaries. An example of a portion of an original text, its romanized form, and its translation is included in Appendix C, along with examples of the renditions in Chinese of some technical terms. A draft of the translation was

then prepared from the original character text, with the aid of the above-mentioned romanized text where necessary. The draft was then compared again with the original character text for accuracy before final editing. Save for one exception (the first translation in this series, of the paper by Zhang Gin-ru which begins on p. 1, was very kindly reviewed by Dr. Andrew Sun), the translations represent entirely my own work.

Although a number of dictionaries were used in preparing the translations (a list is given in Appendix D), for the large majority of unfamiliar characters the first recourse was to the two Chinese — Russian dictionaries published in the USSR (the Kitaisko — Russkii Slovar' and the Kratkii Kitaisko — Russkii Slovar'), since in my experience, the arrangement of characters in these dictionaries is especially convenient for locating a desired character. (The characters are arranged according to a systematic analysis of their morphology in such a way that within limits, and with some practice, these dictionaries can be used in much the same way as an alphabetized dictionary; in contrast, in almost all other Chinese dictionaries, if the pronunciation of a character is not known, an index of the characters, usually based on the 214 Chinese radicals, must first be consulted before locating the character in the body of the dictionary.)

The continuing program of simplification of characters that has taken place in Mainland China in recent years, although having the result that the form employed for some characters was dependent on the year of publication of the particular paper (the same being true for the various dictionaries), proved to be only a minor problem, since, when the need arose, conversion lists (e.g., the "Jianhuazi Zongbiao Jianzi," Peking 1965) were available.

The romanized form of the names of Chinese authors is that used in the abstracts of the original papers (i.e., Wade — Giles or Pinyin); otherwise, the Pinyin system (see Appendix C) has been used. The usual Chinese custom of placing the surname first has been followed, both in the headings for the papers (see Appendix E for originals and abstracts) and in the citations.

The present volume, in more ways than one, owes its existence to Professor Y. R. Chao (Chao Yuen-ren), for it was he who was largely responsible for the development some 40 years ago of the above-mentioned National Romanization System, with which Chinese can readily be written out on an ordinary typewriter (see Appendix C). (As noted by Premier Chou En-lai in "Reform of the Chinese Written Language" [Foreign Language Press, Peking, 1965], this system of romanization in turn formed in part the basis of the Pinyin romanization system, which was established as the standard in Mainland China in 1957.) Moreover, it was Professor Chao's daughter, Dr. R. C. Pian, who conducted the courses in Chinese at Harvard University which I audited in the academic years 1963-65, and which led in time to the present undertaking.

As in previous undertakings of this nature, I am greatly indebted to my wife, Sibylle Jahrreiss Barlow, for preparing the manuscript, and particularly for compiling an expanded page index of the characters in the Kratkii Kitaisko — Russkii Slovar' (available on request), as well as for compiling a conversion table of Pinyin — Cyrillic (Russian) equivalents for use with the Kitaisko — Russkii Slovar'. (Such a table has since become commercially available; see Appendix D.)

The present collection, the year of completion of which coincides with the twentieth anniversary of the establishment of the Peoples' Republic of China, is presented with the express hope that, by making more widely available some of the reports on brain research in China in recent years, a contribution to international understanding as well as scientific exchange may result.

John S. Barlow, M. D.

Department of Neurology
Massachusetts General Hospital
and Harvard University Medical School
Boston, Massachusetts
December, 1969

CONTENTS

INTERACTION OF EVOKED CORTICAL POTENTIALS IN THE RABBIT*

Zhang Gin-ru

Department of Physiology, First Medical College of Shanghai, Shanghai

There have previously been several studies of the interaction of evoked cortical potentials. When paired electrical stimuli of differing intervals were applied to the superficial radial nerve, it was apparent that the evoked cortical potentials of the sensorimotor area exhibited an absolute and a relative inexcitability. With comparatively light anesthesia, there was also apparent an afterdischarge corresponding to an oscillatory decrease in the excitability [1]. In the monkey and in the cat, if direct cortical stimulation were used as the conditioning stimuli and brief sounds were used as the test stimuli and the time course of changes of evoked potentials from the auditory area were examined, a post-excitatory depression as well as oscillatory excitability changes corresponding to the afterdischarge were also seen, such that corresponding to the positive waves of the afterdischarge, cortical excitability decreased [2]. Under the latter experimental conditions, although the mutual interaction of conditioning and test stimuli which results if subcortical afferent systems are employed is avoided, direct cortical stimulation is far from a normal means of activation of the cortex. Consequently, we have presented conditioning and test stimuli to the two somatosensory nerves of different somatic levels, respectively, so that the above-mentioned experimental deficiency could be overcome to some degree, since the extent of the mutual interference of their afferent pathways subcortically is very small. Moreover, utilizing nerves of different somatic levels facilitates the investigation of the spatial aspects of cortical excitability changes as well as features of changes of cortical excitability evoked by different afferent nerves, and similar questions. This paper reports some observations obtained by using this method in rabbits.

METHODS

The experiments were carried out on 32 rabbits, using a chloralose (50 mg/kg) and urethane (500 mg/kg) mixture injected intravenously once; subsequently, small amounts of the anesthetic were injected every one to two hours in order to maintain the anesthesia at a particular level. Stimuli were obtained from a rectangular pulse generator, the stimulator being able to deliver two, or three, separate stimuli. The pulses, of 0.5-0.8 msec width, were passed through an isolation transformer and then applied to the nerves. Conditioning stimuli were usually applied to the inferior orbital nerve, the test stimuli as a rule to the superficial radial nerve; on occasion other sensory nerves were used for stimulation. For recording evoked potentials from the sensorimotor region of the cortex contralateral to the nerve stimulated, the

*Acta Physiologica Sinica, 26(2): 165-171 (1963).

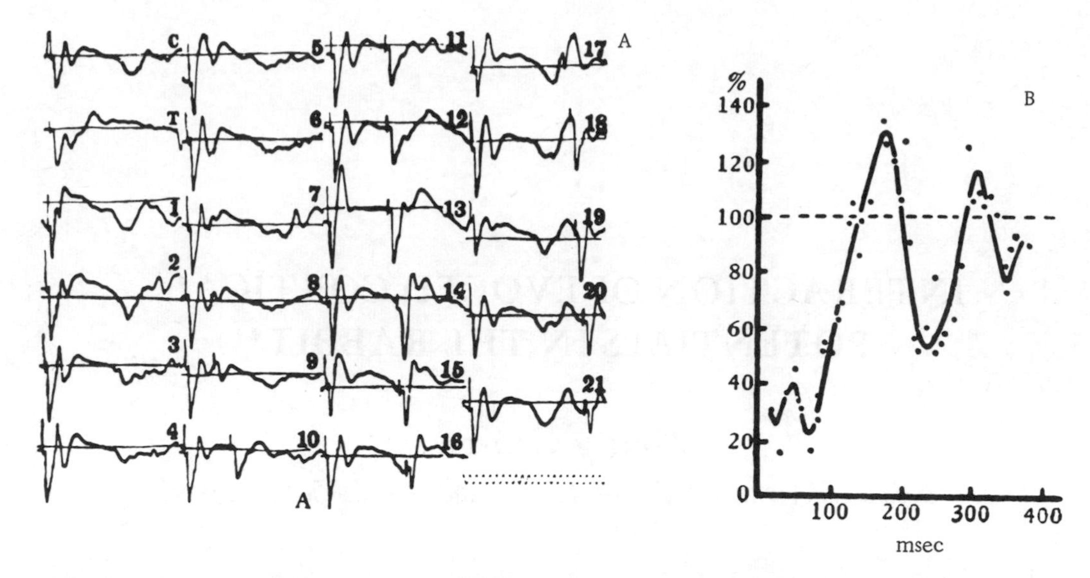

Fig. 1. Temporal aspect of the interaction of evoked potentials. A: Records of electrical potentials. C is the conditioning response; T, the test response; 1-21 are records of the effect of conditioning responses on test responses. Time marker 20 msec. B: Curve of recovery of excitability, derived in part from A. Ordinate: amplitude of the positive wave of the primary component of the test response in percent. Abscissa: time interval between the conditioning and the test stimuli.

exposed cortex was covered with a small amount of mineral oil so as to prevent drying. For the recording electrodes, 0.2-mm-diameter wires, insulated except at the tips with lacquer, were used. The indifferent electrode was placed on the cut edge of the scalp. The amplifiers were condenser-coupled, the time constant always being 0.1 sec. The potentials were photographed from a cathode ray oscilloscope.

RESULTS

1. Temporal Aspects of the Interaction of the Evoked Potentials

Using cortical potentials evoked by stimulation of the inferior orbital nerve as the conditioning response, and cortical potentials evoked by stimulation of the radial nerve as the test response, and recording from the center of the area of the electrical potentials of the test response, it was apparent that during the primary component of the conditioning response, the test response was clearly diminished, exhibiting a post-excitatory depression (Fig. 1A:1-9). During the first positive wave of the afterdischarge of the conditioning response, the test response increased, showing that the excitability was increased (Fig. 1A:13-14); but following recovery from the first positive wave, the test response diminished, showing that the excitability had decreased (Fig. 1A:16-17). During the second positive wave of the afterdischarge, an excitability change similar to that during the first positive wave appeared, but the extent was relatively less (Fig. 1A:19-21). In order to clarify further the time course of the excitability change, the amplitudes of the positive wave of the test responses were calculated as percentages of the control value; these values were then plotted against the time intervals between the conditioning and the test stimuli, yielding the excitability curve (Fig. 1B). From the curve it can also be seen that up to 140 msec after the presentation of the conditioning stimuli, i.e., during the period of the primary component of the conditioning response, the excitability

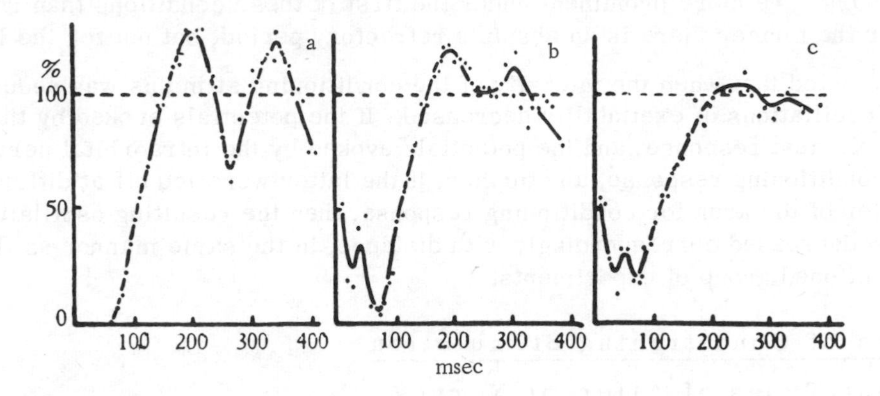

Fig. 2. Spatial summation for evoked potentials. The coordinates for the excitability curve are the same as in Fig. 1. For the conditioning response, potentials evoked by the inferior orbital nerve were used in each instance. For the test response, potentials evoked by the facial nerve were used in a, the superficial radial nerve in b, and the biceps femoris nerve in c. The recording electrode was invariably placed at the center of the area giving the test response. For further explanation, see text.

diminishes, showing a relative refractory period, but there is no absolute refractory period. During the period of the corresponding afterdischarge, the excitability is oscillatory. This result is the same as that of an earlier worker [2], but the amplitude of the oscillations of excitability in the present experiments is somewhat smaller. During the post-excitatory depression that appears during the period of the primary component of the conditioning response, there are two waves of depression which are closely related to the positive potential changes of the primary response. From Fig. 1A it is apparent that the primary component of the conditioning response also possesses two positive waves of potential changes: following the first positive wave of the primary response, the post-excitatory depression is intensified (Fig. 1A:1-2), and following the second positive wave of the primary response, the post-excitatory depression is again intensified (Fig. 1A:6-8). The post-excitatory depression thus has a double oscillation of intensity, and is similar in distribution in the motor area and in other cortical areas not showing the afterdischarge.

If the intensity of the conditioning stimulus is increased, then the post-excitatory depression is enhanced and lengthened and the oscillations in the excitability cycle are also enhanced.

2. The Spatial Factor of the Mutual Interaction of Evoked Potentials

In this group of experiments, the potentials evoked by stimulation of the infraorbital nerve were again used as the conditioning response, but the potentials evoked by the facial nerve, the radial nerve, or the sural nerve were used as the test response. During the observations, the recording electrode was placed at the center of the corresponding area of the cortex. When the potentials evoked by the facial nerve were used as the test response, the center of the areas for the conditioning and the test responses were almost coincident; when the potentials evoked by the radial nerve were used as the test response, the distance separating the two central areas was 2.0 mm, and if the potentials evoked by the sural nerve were used as the test response, the separation between the two central areas increased to 3.5 mm, approximately. The results show that the oscillations in excitability (Fig. 2) resulting from the post-excitatory depression and the afterdischarge are more prominent under the first of these conditions than under the latter two, since for the former there is an absolute refractory period, but not for the latter two.

We also noted that when the intensity of the conditioning stimulus was reduced, the spatial extent of the oscillations of excitability decreased. If the potentials evoked by the radial nerve were used as the test response, and the potentials evoked by the infraorbital nerve used as before for the conditioning response, and further, if the latter were led off at different distances from the center of the area for conditioning response, then the resulting oscillations in the excitability also decreased correspondingly with distance, in the same manner as the results in the above-mentioned group of experiments.

3. Results of Conditioning Stimulation of Different Types of Afferent Nerves

In this group of experiments, the results of potentials evoked by stimulation of the twelfth intercostal nerve or the infraorbital nerve used as the conditioning response were compared initially, using the evoked potentials from the radial nerve as the test response. When the former was used as the conditioning response, the post-excitatory depression was much weaker than for the latter and there was, moreover, no oscillation of the excitability resulting from an afterdischarge. It is known at present that sensory afferents originating from the skin of the trunk do not project strongly to the cerebral cortex, and hence the results of using stimulation of the twelfth intercostal nerve as the conditioning stimulus are understandable.

Next, we also compared the results of using the potentials evoked from the inferior dental nerve (generally considered to subserve pain) and the inferior orbital nerve (a cutaneous sensory nerve) as the conditioned response, and the potentials evoked by the radial nerve as the test response, as before. The experiments showed that the two did not differ appreciably. If the potentials evoked by the nerve of the biceps femoris muscle (a proprioceptive sensory afferent nerve) and the sural nerve (a cutaneous sensory nerve) were selected as the conditioning responses, then the results of the two also did not differ significantly. From this it is evident that the oscillations in cortical excitability resulting from potentials evoked by different types of sensory nerves are similar.

4. Effect of Depth of Anesthesia on the Recovery of Excitability

In this group of experiments, the potentials evoked from the infraorbital nerve were again used as the conditioned response, and the evoked potentials from the radial nerve were used as the test response. Initially, the excitability curve was determined under the usual anesthetic conditions (Fig. 3a); the depth of anesthesia was then gradually increased and the corresponding changes in the curve were observed. It was seen that as the depth of anesthesia was increased, the post-excitatory depression gradually diminished, and its duration gradually increased; the oscillation in excitability resulting from the afterdischarge gradually disappeared (Fig. 3b and 3c). Under conditions of deep anesthesia, the post-excitatory depression could extend to 0.8 sec, approximately (Fig. 3d), and at the same time the double fluctuation portion of the post-excitatory depression become more clearly differentiated. It is thus evident that the mechanisms for formation of these two waves of reinforcement are different.

5. Effect of Strychnine and Curare on the Recovery of Excitability

In this group of experiments, 2 × 2 mm pieces of filter paper soaked in 3% strychnine solution were placed on the surface of the cortex, and the effect of strychnine on the recovery of

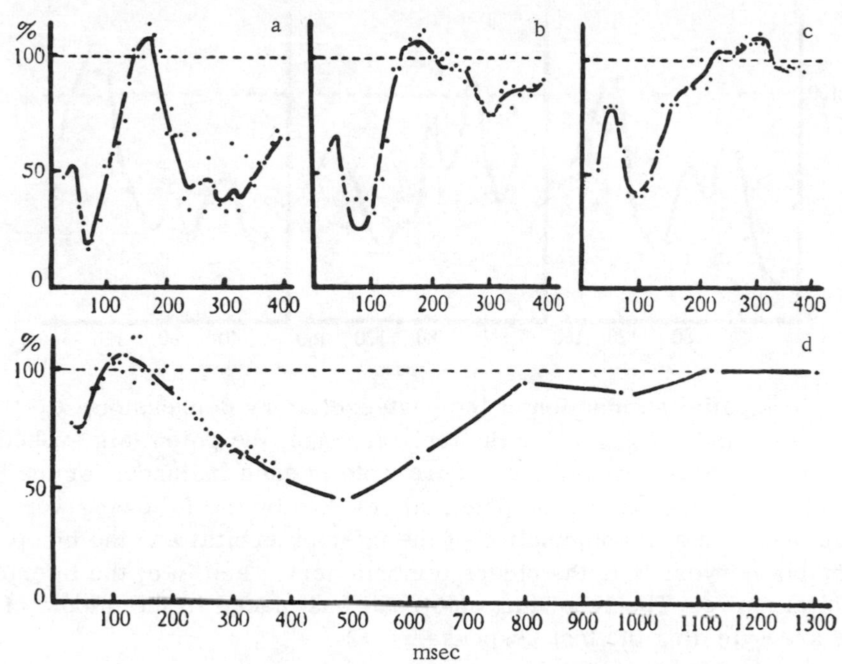

Fig. 3. Effect of depth of anesthesia. Coordinates the same as in Fig. 1. The curve in a is the excitability curve for the usual conditions of anesthesia; b is the result following injection of a 3-ml mixture of the anesthetic, of which each ml of the mixture contained 10 mg of chloralose and 100 mg of urethane; c and d are the results following two successive injections of 3 ml of the anesthetic.

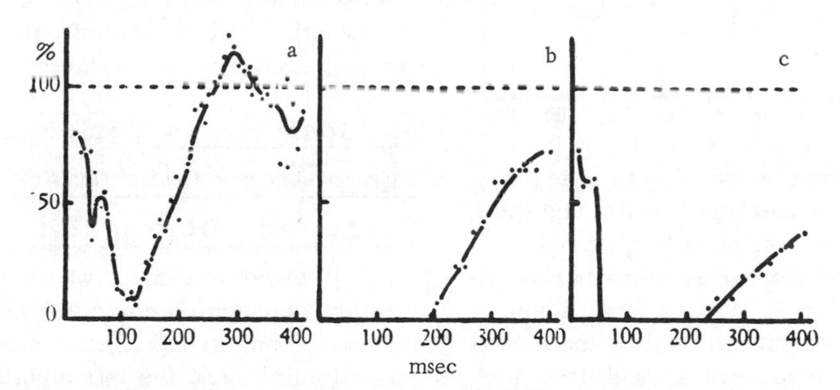

Fig. 4. Effect of strychnine. Coordinates as in Fig. 1. The control curve is shown in a; b shows the effect of 3% strychnine; c is the result following injection of a 6-ml dose of the anesthetic mixture.

excitability was observed from the potential changes led off from a small hole in the center of the filter paper. It was seen that with strychnine the effect of the after-potentials was clearly increased; the post-excitatory depression was clearly enhanced and prolonged, and an absolute refractory period was even present (Fig. 4b). If at this time the depth of anesthesia were increased as a way of enhancing the separation of the two fluctuations in the post-excitatory de-

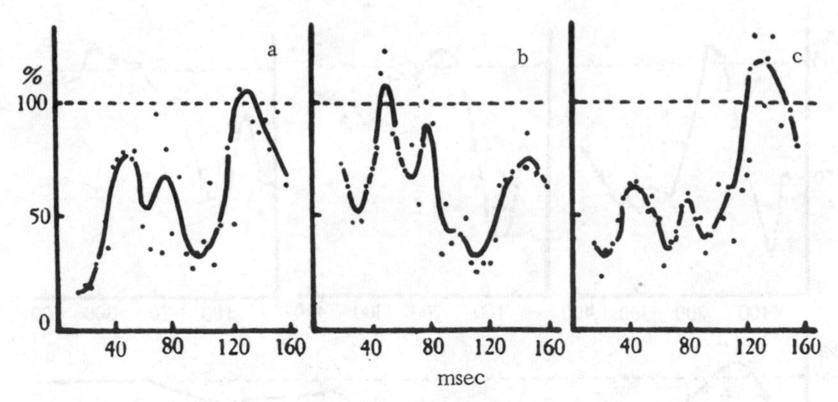

Fig. 5. Spatial summation of the post-excitatory depression. Coordinates as in Fig. 1. For the test response, the potentials evoked by the superficial radial nerve were used in each instance; for the conditioning responses, the potentials evoked by the following were employed: in a, a combination of the inferior orbital and the biceps femoris nerves; in b, the biceps femoris nerve; and in c, the inferior orbital nerve. The recording electrode was placed in the middle of the area yielding the test response.

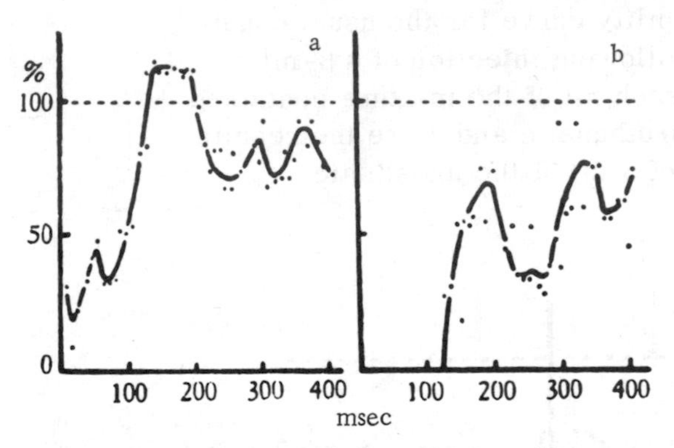

Fig. 6. Comparison of the effects of test responses from the superficial radial and the biceps femoris nerves; conditioning responses from the inferior alveolar nerve.Coordinates as in Fig. 1. For the conditioning response, the potentials evoked by the inferior orbital nerve were always used; in a, the test responses were potentials evoked by the superficial radial nerve; in b, the control responses were the potentials evoked by the same inferior orbital nerve. The recording electrode was placed in the center of the area giving the evoked potentials for the superficial radial nerve.

pression, then it was apparent that strychnine had an effect only on the second increase in depression (Fig. 4c). These observations are in agreement with the results of previous studies [2].

Curare and strychnine possess similar effects on cortical neurons [3]; we found that the two also had similar effects on the fluctuations of recovery of excitability.

6. Spatial and Temporal Summation of the Post-Excitatory Depression

In order to clarify whether the post-excitatory depression could exhibit spatial summation, in one group of experiments simultaneous stimulation of the infraorbital and the sural nerves was used for the conditioning stimulus, and the potentials evoked by the superficial radial nerve served as the test response as before. Initially, the post-excitatory depression was determined separately for the two conditioning responses (Fig. 5b and c). Then, a method was devised so that the two conditioning responses appeared simultaneously at the cortex, following which the changes in the test responses were examined for different time intervals, in order to ascertain whether or not the phenomenon of spatial summation occurred at these times. In order to insure that the two conditioned reponses appeared simultaneously at the cortex, the latent periods for the evoked potentials for the sural and for the infraorbital nerves, respectively, were measured initially; the result for the former was 15 msec, and for the latter, 9 msec. Consequently, the stimulus applied to the sural nerve was made to precede that to the infraorbital nerve by 6 msec, so that the two conditioned responses appeared at the cortex just at the same time. When the two conditioning responses appeared simultaneously at the cortex, the initial portion of post-excitatory depression changed, becoming larger than the effect for each of the two separately (Fig. 5a). For example, in Fig. 5b, in the 20- to 30-msec interval following stimulation of the sural nerve, the value of the inhibition is 30-40%. In Fig. 5c, the value is approximately 80-85%. It is apparent that for a period the two types of inhibitory responses have a combined effect, i.e., the post-excitatory depression of the two appears to exhibit partial spatial summation. Under the conditions of their joint effect, the later part of the post-excitatory depression becomes relatively complex and unstable (Fig. 5a), and consequently, further analysis was difficult.

In order to clarify whether the post-excitatory depression exhibited temporal summation, in another series of experiments we compared the time course of electrical responses evoked from the infraorbital nerve with those from the superficial radial nerve (Fig. 6a), as well as the time course under conditions in which the potentials evoked from the infraorbital nerve were used for both the conditioning and the test responses (Fig. 6b). It is apparent that under the latter conditions, the post-excitatory depression is much longer and more intense than for the former conditions; there is even an absolute refractory period. From this it is evident that in the trials of combining two stimuli applied to the infraorbital nerve, spatial summation of the post-excitatory depression is impossible with respect to the test response obtained at that time from the superficial radial nerve, because at the time the test response from the superficial radial nerve is gradually released from the effect of the post-excitatory inhibition of the preceding infraorbital nerve stimulation, the inhibitory response from the following stimulus to the infraorbital nerve cannot yet have developed, for the reason that the second response has not yet been removed from the post-excitatory depression of the preceding response from the same afferent system.

7. Results of Analysis of the Effects of Conditioning and Test Stimuli to the Two Sides of the Body

When test stimuli were presented to the right superficial radial nerve and the test responses were evoked from the center of the respective area of the cortex on the left side and the recovery of excitability was observed, it was seen that the results were the same for conditioning stimuli delivered to the left or the right infraorbital nerves, that there was no difference in the recovery of excitability. This is because the afferent projection of the infraorbital nerves to the cortex is bilateral. In the same animal, the results for conditioning stimuli differentially presented to the sural nerve on the two sides were different: the post-excitatory depression of the cortex on the left side resulting from stimulation of the left sural nerve was much weaker than that for stimulation of the right side. This is because the afferent projection of the sural nerve to the cortex is crossed, and its effect on the ipsilateral cortex is accomplished by means of indirect pathways.

8. The Recovery of Excitability of the Secondary Somatosensory Cortical Area

All of the above-mentioned results were obtained from the primary somatosensory cortex. In one group of experiments, we recorded from an electrode placed on the secondary somatosensory area, and as before analyzed conditioning and test responses, using the evoked potentials from the infraorbital and the superficial radial nerves, respectively. It was seen that the recovery of excitability was fundamentally similar to that for the primary somatosensory area, but the post-excitatory depression was somewhat stronger and there was an absolute refractory period. This is because of the phenomenon of overlap of the somatosensory projections in the secondary somatosensory region; the central areas for the test and the conditioning responses are very close, and hence their mutual interaction is necessarily prominent.

DISCUSSION

From the similarity of the effects of potentials evoked by stimulation of sensory nerves at different segmental levels, it is clear that their projections to the somatosensory cortex show the phenomenon of overlap. If the evoked potentials for a given level reflect sensory activity, then from the present experimental results it was apparent that the sensory excitation transmitted from different locations on the body can mutually interfere at the level of the cerebral cortex, the excitatory sensory activity of a preceding one being capable of altering that of a subsequent one for a particular interval of time. This effect can be explained by the mechanism of the mutual partial interference for different sensory points of the body. It is apparent that sensory stimuli that are strongly projected to the cortex (for example, sensory stimulation of the extreme points on the body, such as the fingers, toes, and the lips) interfere relatively strongly with other somatosensory points, but the stimulation of some points is projected relatively weakly to the cortex (for example, sensory stimulation of areas on the trunk) and therefore they do not have a prominent interference effect. At the same time, the mutual interference between two neighboring sensory projections at the cortical area is relatively strong. There are no clear differences concerning mutual interference with respect to the modality of the sensory afferent (pain, cutaneous sensation, or muscle and proprioceptive sensation).

The degree of oscillation in excitability resulting from the afterdischarge which was seen in the present experiments was smaller than in the results of an earlier study [2], in particular, the degree of increase of excitability was not large; these differences can perhaps be related to the following factors: (1) the experimental animals were different; for the earlier results cats and monkeys were utilized, whereas in the present experiments, rabbits were employed; (2) the manner for evoking the conditioning responses was different; in the present experiments the conditioning responses were evoked by stimulation of the peripheral sensory nerves rather than being evoked by stimuli delivered directly to the cortex; (3) the center of the area of, and the interelectrode separation for the leads for, the conditioning responses were different; in the present experiments the distances were all somewhat larger.

SUMMARY

Using the electrical potentials evoked at the cortex by stimulation of nerves of different segmental levels as an index, the course of recovery of cortical excitability has been observed in rabbits. Following activation of the cortex by a single afferent impulse, a period of unresponsiveness is immediately apparent, following which there is a periodic alteration in the excitability which corresponds to the afterdischarge. This fluctuation in excitability has a definite spatial distribution: in the center of the responsive area, the oscillation is strong and there is an absolute refractory period; in the peripheral region, the fluctuation gradually diminishes and there is no absolute refractory period. The degree of fluctuation of the excitability and the size of the spatial distribution are determined by the strength of the cortical projection of the respective sensory nerve, and are not related to the sensory modality. With increased depth of anesthesia, the fluctuation in excitability diminishes, but its duration is clearly increased.

The author wishes especially to thank Professor H. -T. Chang, of the Institute of Physiology, Chinese Academy of Sciences, for his guidance in the present work.

REFERENCES

1. Jarcho, L. W., Excitability of cortical afferent systems during barbiturate anesthesia, J. Neurophysiol., 12:447-457 (1949).
2. Chang, H.-T., Changes in excitability of the cerebral cortex following a single electric shock applied to the cortical surface, J. Neurophysiol., 14:95-111 (1951).
3. Chang, H.-T., Similarity in action between curare and strychnine on cortical neurons, J. Neurophysiol., 16:221-233 (1953).

CORTICAL REPETITIVE RESPONSES ELICITED BY A SINGLE CONTRALATERAL STIMULUS*

Fan Shih-fang and Shen Ke-fei

Institute of Physiology, Chinese Academy of Sciences, Shanghai

In response to direct cortical stimulation with single stimuli in unanesthetized animals, a train of surface-negative repetitive responses with a frequency of 10-20/sec can be evoked; these can be considered to result from excitation of closed circuits of the cortex and the medial thalamus, the result of periodic impingement of neural impulses on the apical dendrites of pyramidal neurons [1]. In further work, we also observed that after the intensity of stimulation was increased, the contralateral cortex also showed the same type of repetitive response; although the latter was not synchronous with that of the ipsilateral side, it did not disappear upon sectioning the corpus callosum.

METHODS

The experiments were carried out on rabbits. Usually in an experiment, the cerebral cortex was first exposed, under ether anesthesia. In a part of the experiments, the visual cortex and hippocampal gyrus were removed by suction so as to expose the thalamus. During this operation, care was taken not to damage the connections of the corpus callosum with the motor area of the two cerebral hemispheres. Upon completion of the surgical operation, ether was discontinued, and the animals were paralyzed by injection of curarine.

The changes of potential were recorded by means of a dual beam cathode ray oscilloscope; the experimental method was otherwise basically the same as that previously used [1], and hence will not be repeated.

RESULTS

General Appearance of the Responses

Upon direct cortical stimulation of the cerebral cortex of an animal on one side and recording from a unipolar electrode on the corresponding point of the contralateral side, an early positive late negative potential change is recorded, in the main, which is evoked by nerve impulse activity that is transmitted via the corpus callosum [2, 3]. Following this biphasic electrical potential, an interval with a relatively prolonged surface-negative potential also sometimes appears. In the unanesthetized animal, if the stimulus intensity is relatively strong, then following these potential changes a train of prominent repetitive surface-negative responses

*Acta Physiologica Sinica, 25(2):114-118 (1962).

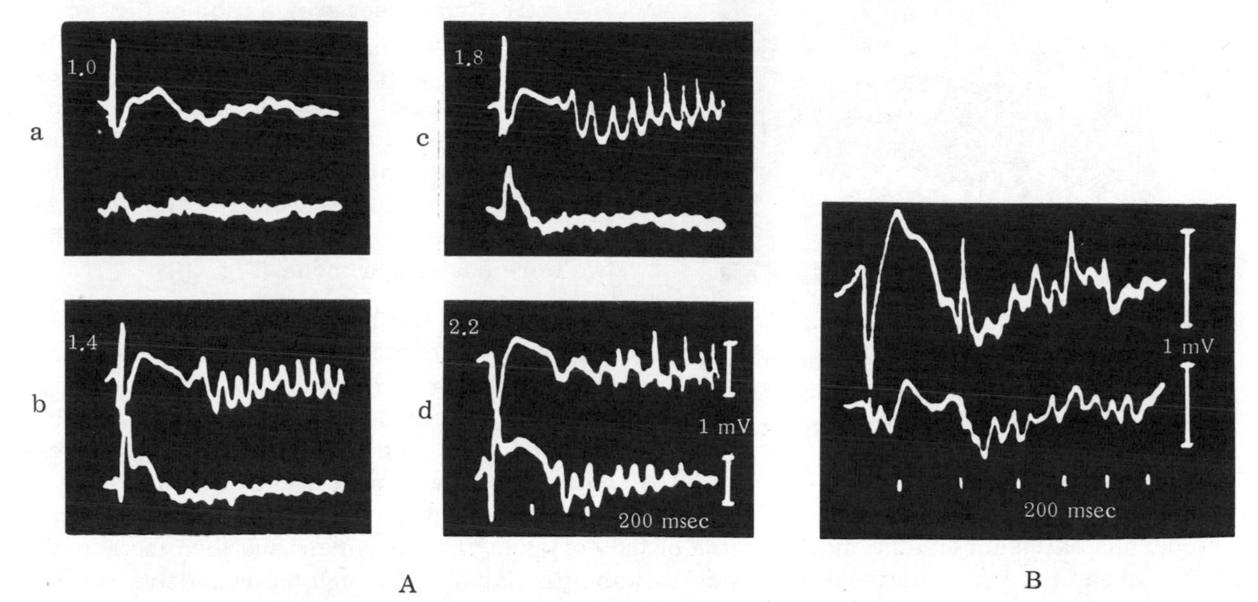

Fig. 1. Responses evoked by direct cortical stimulation with single electrical stimuli. The upper trace in each photograph is the result obtained from a unipolar electrode 1-2 mm distant from the stimulating electrode, the lower trace is the result obtained from the symmetrical point on the contralateral hemisphere. A. Results obtained from the sensorimotor area, showing that for a relatively weak stimulus, the repetitive responses are evoked only on the ipsilateral side (b-c), whereas, when the stimulus strength is increased, the repetitive responses are also evoked on the contralateral side (d). The numbers in the upper left corner of each photograph are the stimulus intensities, in relative units. B. Results obtained from the visual area.

also appears. Figure 1 is an example of the result. Figure 1A was obtained from the sensorimotor area with a relatively weak stimulus; the repetitive response appears only on the ipsilateral side (Fig. 1A; b, c). With a somewhat stronger stimulus, the repetitive response also appears on the contralateral cortex (Fig. 1A; d). Although the waveform and the frequency of the repetitive response appearing on the contralateral side are similar to those for the ipsilateral side, the two sides are not synchronous. This type of repetitive response upon stimulation of the sensorimotor area of one side can be obtained over almost all of the sensorimotor cortical area of the contralateral side. Figure 1B is an example of results obtained for the visual area; in the main it is similar to the response obtained for the sensorimotor area, although it is not as prominent, and the intensity of stimulation needed was higher.

The Corpus Callosum as a Possible Intermediary in Contralateral Cortical Repetitive Responses to Stimulation of the Ipsilateral Cortex

By means of recordings from small steel needle electrodes inserted into the corpus callosum, repetitive potential changes could be found in the corpus callosum at the same time that repetitive cortical responses were being evoked by stimulation of the contralateral cortex, although they were not synchronous with either the responses of the ipsilateral cortex or with those of the contralateral cortex. This indicates that the contralateral repetitive response cannot be evoked by periodic nerve impulse activity transmitted via the corpus callosum, but instead is perhaps excited together with the activity evoked in the ipsilateral cortico-thalamic circuits mentioned. In order to verify this point, we carried out experiments to answer the following two questions:

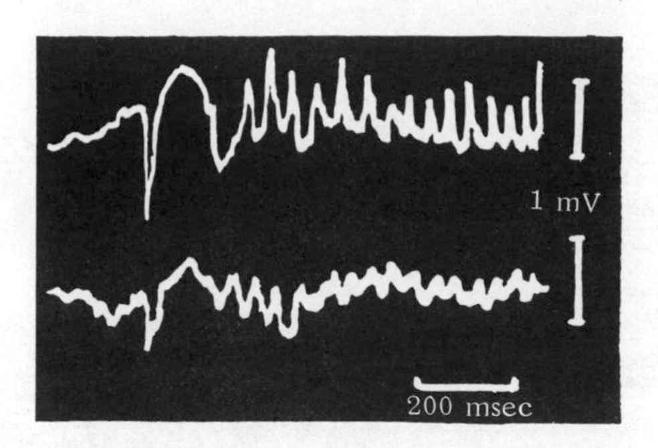

Fig. 2. Bilateral cortical responses resulting from direct stimulation of the corpus callosum by means of a pair of needle electrodes. Note that the repetitive responses of the two sides are not synchronous.

(1) Can direct stimulation of the corpus callosum evoke repetitive cortical responses ? We inserted a pair of steel needle electrodes vertically between the two hemispheres into the corpus callosum for stimulation, and obtained repetitive responses from the cortex on both sides; their waveforms were similar to those evoked by direct cortical stimulation, but the two were not synchronous (Fig. 2).

(2) Can the repetitive response still be evoked in the cortex on one side by nerve impulses passing through the corpus callosum, after removal of the thalamus on that side by suction ? The medial part of the thalamus on one side (side A) was removed by suction, care being taken not to injure the interconnections between the corpus callosum and the motor region of the cortex on the two sides, and then the sensorimotor area of the contralateral side (side B) was stimulated. Although the repetitive response was still clearly apparent from the cortex on the B side (Fig. 3a, upper trace), the cortex on the A side showed only a relatively simple biphasic potential (Fig. 3a, lower trace). If the cortex on the A side were stimulated at this time, then a repetitive response was still not evoked from the cortex on the A side (Fig. 3b, lower trace); but a repetitive response was nevertheless clearly evoked from the cortex on the B side (Fig. 3b, upper trace).

The above two types of results clearly show that stimulation of the cortex on one side can pass through the corpus callosum and evoke activity in the contralateral cortico-thalamic circuits.

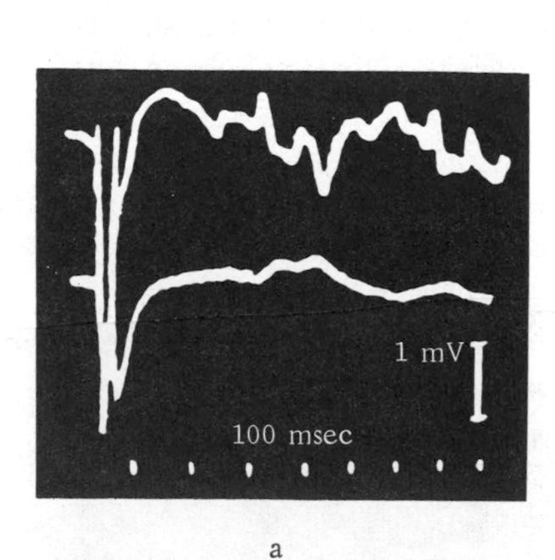

a

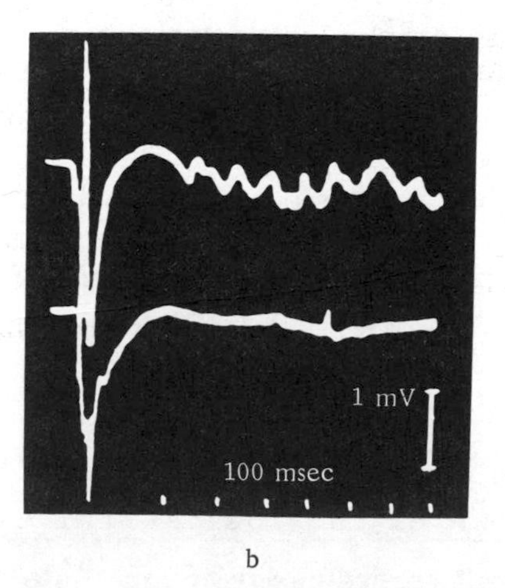

b

Fig. 3. Responses obtained from the cortex on the two sides for cortical stimulation after unilateral removal of the medial part of the thalamus by suction. (a) Stimulation of the cortex on the side of the intact thalamus evokes a repetitive response only from the ipsilateral cortex (upper trace); there are no repetitive responses from the contralateral cortex (lower trace); (b) Stimulation of the cortex on the side of thalamic removal by suction evokes repetitive responses from the cortex of the contralateral side only (upper trace); there are no repetitive responses from the ipsilateral cortex (lower trace).

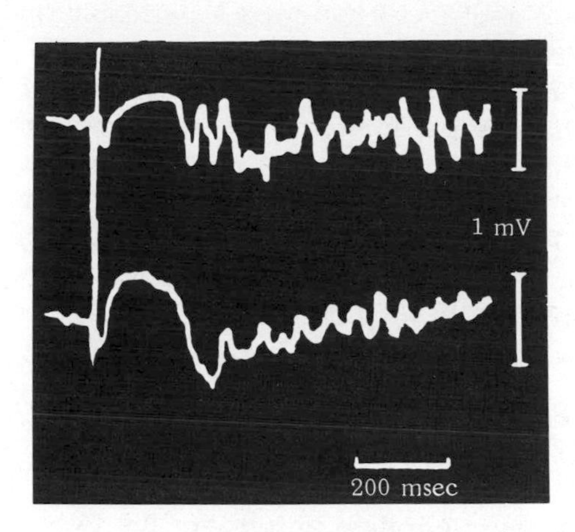

Fig. 4. Persistence of contralateral cortical repetitive responses by cortical stimulation, following sectioning of the corpus callosum. The upper trace is the result recorded from a point 2 mm from the stimulating electrode. The lower trace is the result from the symmetrical point on the contralateral cortex. The recordings are from the sensorimotor area.

Capability of Ipsilateral Cortical Stimulation to Evoke a Repetitive Response in the Contralateral Cortex via the Thalamus

As a further step, we first cut the corpus callosum, and then stimulated the cortex on one side, with the result that a repetitive response could usually still be obtained from the contralateral cortex. This type of result was, as before, easier to obtain from the sensorimotor area than from the visual area. Figure 4 is an example of the results obtained from the sensorimotor area; the upper trace is from the ipsilateral cortex, the lower trace from the contralateral side. The amplitude and duration of the repetitive response from the contralateral cortex was invariably very much reduced in comparison with that obtained prior to section of the corpus callosum. At this time, if a midline transection of the thalamus were then made, the repetitive response completely disappeared from the contralateral cortex. These results show that stimulation of the cortex on one side can evoke a repetitive response from the contralateral cortex via the thalamus.

DISCUSSION

Four years ago, using curarized animals, we observed that upon direct cortical stimulation by single stimuli, a surface-negative repetitive electrical potential was usually recorded from the ipsilateral cortex. Further, we also observed that following stronger stimulation, a repetitive response of a similar character could also appear from the contralateral cortex. At that time, we considered that this response resulted from excitation of some kind of closed circuit between the cerebral cortex and the medial part of the thalamus. Repetitive responses can also be evoked from the cerebral cortex by single stimuli applied to the caudate nucleus [4, 5]. This type of response is very similar to that evoked by cortical stimulation in the following respects (compare [4, 6-8]): (1) the form of the response; (2) relatively greater prominence in the sensorimotor area; (3) responses evoked bilaterally from unilateral stimulation; (4) great sensitivity to ether and pentobarbital anesthesia; (5) ready elimination by localized application of procaine to the cortex. According to our results, however, the medial part of the thalamus, for example, the nucleus dorsalis medialis, is necessary for the appearance of repetitive responses to cortical stimulation, but according to the report of Buchwald et al. [6], only the nucleus ventralis anterior of the thalamus is necessary for the appearance of repetitive responses to stimulation of the caudate nucleus. Consequently, two possibilities exist: (1) the two responses, although of similar character, are different in the mechanisms of their formation; (2) the mechanisms of production of the two have a common external source: both are related to activity in the nucleus ventralis anterior, but the latter nucleus is included in the aforementioned closed circuits between the cortex and the medial part of the thalamus. From the many similar characteristics of the two responses, we are inclined toward the latter possibility.

There have already been numerous reports concerning cortical responses evoked by contralateral thalamic stimulation. For example, Jasper [9] described cortical responses in the frontal sensorimotor area that could be evoked bilaterally by unilateral stimulation of the thalamus; this phenomenon could still appear after complete section of the corpus callosum and the anterior and posterior commissures. Kerr and O'Leary [10] also reported that bilateral cortical responses could be evoked in the rabbit by unilateral stimulation of the medial part of the thalamus. The results of the present report further show that a cortical repetitive response can also be evoked via the thalamus, on the contralateral side by ipsilateral stimulation of the cortex. This fact shows that the thalamus has a considerable influence on the relation of inter-hemispheric activity. Jasper and Droogleever-Fortuyn [11] and Garoutte and Aird [12] reported that in patients after section of the corpus callosum and in patients with incomplete development of the latter, the electrical activity of the two hemispheres is still definitely related; this circumstance could quite possibly be the result of the effect of the thalamus between them.

Whether there is a common subcortical pacemaker for the activity of the two hemispheres is still disputed. Jasper [9, 11, 14] and Garoutte and Aird [12, 13] consider that a subcortical pacemaker does exist which controls the electrical activity of the cortex on both sides; but others (for example, Cohn [15]) assume that the circumstance that the electrical activity of the two hemispheres is not completely synchronous is evidence against the existence of such a subcortical pacemaker. The present work does not provide data directly related to this dispute, although from the results of the present report, it is apparent that if there indeed exists a subcortical pacemaker, then its effect is primarily to evoke thalamic activity (as suggested by Garoutte and Aird [12]), so that the activity of the two hemispheres is not completely synchronized.

SUMMARY

Using direct electrical stimulation of the cerebral cortex of the unanesthetized rabbit with single stimuli, a response was evoked from the corresponding area of the contralateral cortex which included a train of repetitive, primarily surface negative, waves of a frequency of 10-20/sec. Direct electrical stimulation of the corpus callosum could also evoke repetitive cortical responses bilaterally, although the two sides were not synchronous. After section of the corpus callosum, this response was still apparent, but after midline section of both the corpus callosum and the thalamus, the response disappeared. After suctioning away the thalamus on one side, but preserving the corpus callosum, the repetitive response was obtained only from the cortex on the side of the intact thalamus, irrespective of which side the cortex was stimulated. From these results it is apparent that stimulation of the cortex on one side can excite cortico-thalamic circuits on the contralateral side, via the corpus callosum and the thalamus, and thus evoke the repetitive responses.

REFERENCES

1. Shen Ke-fei and Fan Shih-fang, Repetitive surface-negative potentials evoked by single stimuli to the cortex (in Chinese), Acta Physiol. Sinica, 22:167-174 (1958).
2. Curtis, H. J., Intercortical connection of corpus callosum as indicated by evoked potentials, J. Neurophysiol., 3:407-413 (1940).
3. Chang, H. T., Cortical response to activity of callosal neurons, J. Neurophysiol., 16:117-131 (1953).
4. Shimamoto, T., and Verzeano, M., Relations between caudate and diffusely projecting thalamic nuclei, J. Neurophysiol., 17:278-288 (1954).
5. Buchwald, N. A., Wieck, H. H., and Wyers, E. J., Effects of stimulation of caudate nucleus on outflow of globus pallidus, Anat. Rec., 133:256 (1959).
6. Buchwald, N. A., Wyers, E. J., Okuma, T., and Heuser, G., The "caudate spindle." I. Electrophysiological properties, Electroenceph. clin. Neurophysiol., 13:509-518 (1961).
7. Heuser, G., Buchwald, N. A., and Wyers, E. J., The "caudate spindle." II. Facilitatory and inhibitory caudate-cortical pathways, Electroenceph. clin. Neurophysiol., 13:519-524 (1961).

8. Buchwald, N. A., Heuser, G., Wyers, E. J., and Lauprecht, C. W., The "caudate spindle," III. Inhibition by high frequency stimulation of subcortical structures, Electroenceph. clin. Neurophysiol., 13:525-530 (1961).
9. Jasper, H. H., Diffuse projection systems: The integrative action of the thalamic reticular system, Electroenceph. clin. Neurophysiol., 1:405-420 (1949).
10. Kerr, F. W. L., and O'Leary, J. L., The thalamic source of cortical recruiting in the rodent, Electroenceph. clin. Neurophysiol., 9:461-476 (1957).
11. Jasper, H. H., and Droogleever-Fortuyn, J., Experimental studies on the functional anatomy of petit mal epilepsy, Res. Publ. Ass. nerv. ment. Dis., 26:272-298 (1947) (cited by Garoutte and Aird [12]).
12. Garoutte, B., and Aird, R. B., Studies on the cortical pacemaker: Synchrony and asychrony of bilaterally recorded alpha and beta activity, Electroenceph. clin. Neurophysiol., 10:259-268 (1958).
13. Ogden, T. E., Aird, R. B., and Garoutte, B. C., The nature of bilateral and synchronous cerebral spiking, Acta Psychiat. Neurol. Scand., 31:273-284 (1956).
14. Jasper, H., Reflections on the spike and wave complex in cortical and centrencephalic systems, Electroenceph. clin. Neurophysiol., 9:379 (1957).
15. Cohn, R., Spike-dome complex in the human electroencephalogram, Arch. Neurol. Psychiat. (Chicago), 71:699-706 (1954).

CORTICAL RESPONSES TO REPETITIVE CONTRALATERAL STIMULATION AFTER SECTIONING OF THE CORPUS CALLOSUM*

Fan Shih-fang and Shen Ke-fei

Institute of Physiology, Chinese Academy of Sciences, Shanghai

After sectioning of the corpus callosum, direct cortical stimulation with single stimuli in the unanesthetized animal can still evoke a train of mainly surface negative responses from the contralateral cortex [1]. This kind of response only appears with relatively intense stimuli, and is not seen with reduced intensities of stimulation. In further work, we observed that if low-frequency repetitive cortical stimulation were used, and if the stimulus intensity were not too great, then electrical potentials could still be evoked from the contralateral cortex following section of the corpus callosum.

METHODS

The experiments were carried out on rabbits paralyzed with curare. Initially, in an experiment, under ether anesthesia, the cortex was exposed and covered with mineral oil to prevent drying. Upon completion of the operation, the ether anesthesia was discontinued, the animals were injected with curarine to paralyze them, and artificial respiration at a rate of 20/min was instituted.

For the stimulating electrodes, a pair of 0.7-mm-diameter steel wires were used, the tips of which were separated by a distance of 1.5 mm. The stimuli were brief pulses of a duration of 0.8 msec, delivered to the stimulating electrodes via an isolation unit. For recording from the cortical surface, electrodes were prepared from 0.7-mm-diameter silver wires, and a small sheet of silver applied to the inner surface of the cut edge of the scalp served as the indifferent electrode. For recording from the thalamus, the electrodes were made from stainless steel needles insulated except for the tip, of which the diameter was 15 μ. For examination of the potential changes, a two-channel cathode ray oscilloscope of our own manufacture and an Ediswan Mark II electroencephalograph were used. The former was equipped with DC amplifiers; when it was necessary to use the latter, the time constant of the amplifiers was generally set at 1 sec.

Upon conclusion of the observations in the experiments in which responses were recorded from the thalamus, a direct current (of 1 mA, for approximately 0.5 min) was passed through the steel microelectrode serving as the anode. After the experiment was concluded, the brain

*Acta Physiologica Sinica, 26(3): 211-217 (1963).

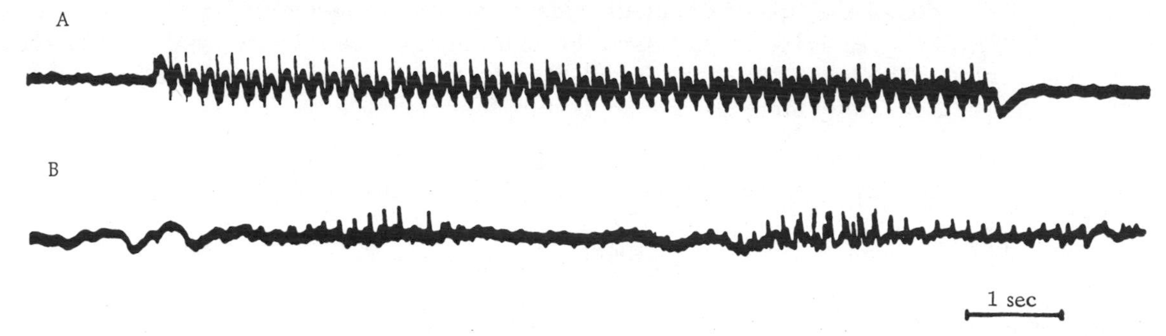

Fig. 1. Electrical responses evoked by repetitive contralateral cortical stimulation before (A) and after (B) sectioning of the corpus callosum. Both the stimulating and the recording electrodes were placed on the sensorimotor area. For detailed explanation, see text.

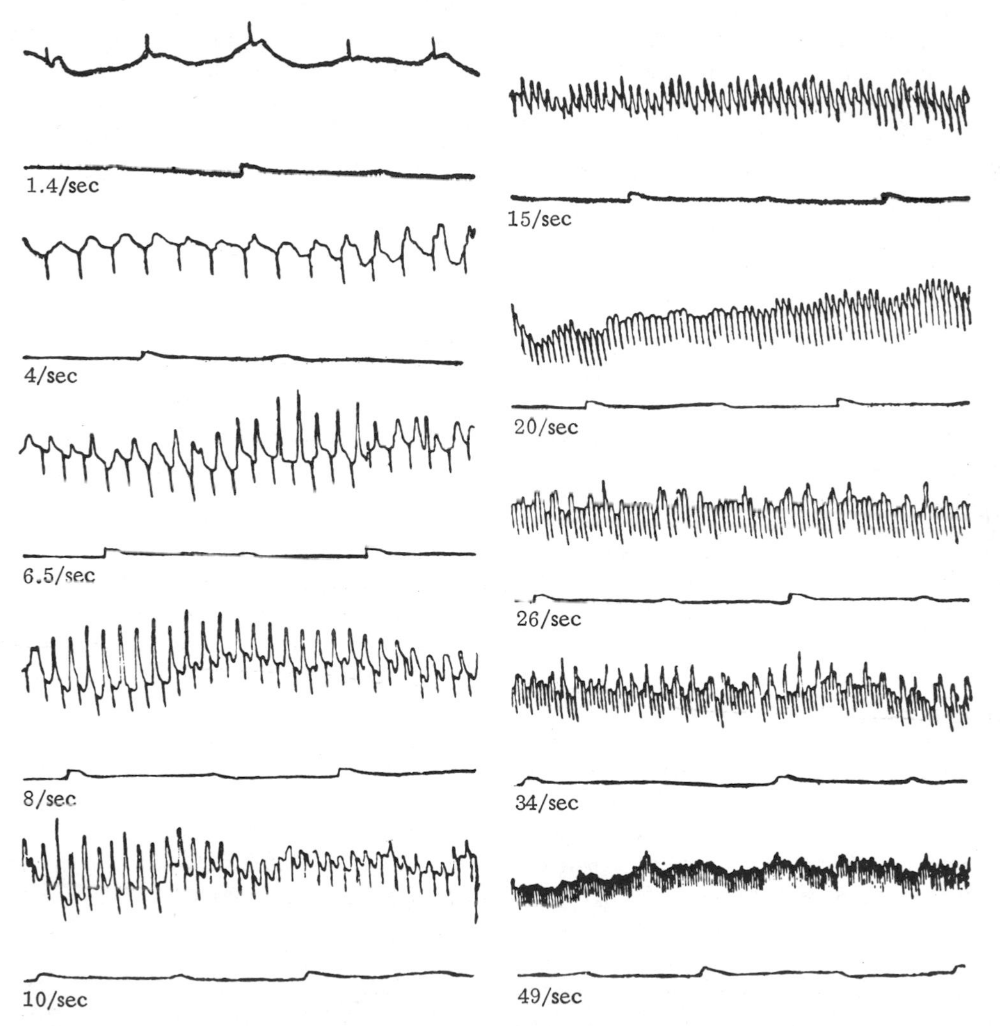

Fig. 2. Electrical responses evoked by contralateral repetitive stimulation at different frequencies, after sectioning of the corpus callosum. Time marker: 1 sec. For detailed explanation, see text.

was removed, fixed in 10% formalin, and after several days, was removed, sectioned, and examined. The locations of the tips of the electrodes were determined with the aid of the atlas of Sawyer et al. [2], the parts of the brain tissue through which current had flowed usually showing a dark brown color. The locations of the tips of the electrodes measured by this method were in agreement with those determined by the potassium ferrocyanide method [3].

The experiments were carried out in a small electrically shielded and thermally insulated chamber, in which the temperature was maintained at approximately 30°C.

RESULTS

General Appearance of the Responses

With direct cortical stimulation using low frequency (8-10/sec) repetitive stimuli, if the stimulus intensity is not too great, following each of the stimuli the contralateral cortex generally shows an early positive late negative potential change that is evoked by nerve impulses transmitted via the corpus callosum (hereafter termed the callosal response) [4, 5]. Following this biphasic potential, in some experiments, a relatively prolonged surface-negative potential also appeared. After the corpus callosum was sectioned, such stimuli no longer evoked the above-mentioned potentials from the contralateral cortex, but after the very first stimulus, a surface-negative response of a relatively long latency accompanied each of the stimuli. The amplitudes of these responses were periodic: at first gradually increasing, then gradually decreasing, and then again gradually increasing, thus repeating the cycle (hereafter abbreviated as the waxing-and-waning response). Figure 1 shows an example of the results for a stimulus frequency of 7.3/sec. Figure 1A shows the response before sectioning the corpus callosum; following each stimulus there is a clear callosal response. Figure 1B shows the response following callosal sectioning; no clear-cut response is evident for initial stimuli, but beginning with the sixth stimulus, surface-negative potentials appear, the amplitudes of which increase, then decrease, even to the point of disappearing, and after an interval of time again gradually increase. By using a faster sweep for recording, it was observed that the latent period for each response exceeded 60 msec. In different experiments and at different times during the same experiment, there were relatively large changes in the latent period, although it was generally greater than 30 msec.

If the frequency of stimulation were decreased to less than 7/sec, the waxing-and-waning response either did not appear, or appeared only after a larger number of stimuli. For stimulus frequencies of 7-10/sec, the response readily appeared, but for stimulus frequencies of 10-20/sec, the response was again relatively difficult to see. For stimulus frequencies of 20/sec, the waxing-and-waning response either did not appear, or appeared as for approximately 8/sec stimuli (Fig. 2).

When the stimuli were applied to the sensorimotor area of the cortex, the waxing-and-waning response could generally be obtained from almost the entire contralateral sensorimotor area. When the visual area of the cortex was stimulated, a similar response could also be evoked from the contralateral visual area, but under the same experimental conditions, for example, with the same animal, the requirements of frequency and intensity of stimulation were more strict than for the sensorimotor area.

The Problem of the Pathway for the Response

In animals in which the corpus callosum had already been sectioned, we further transected the thalamus along the midline; the waxing-and-waning response could then no longer be recorded from the contralateral cortex. Further, while the waxing-and-waning response was being obtained, we inserted a steel recording microelectrode into different parts of the contralateral

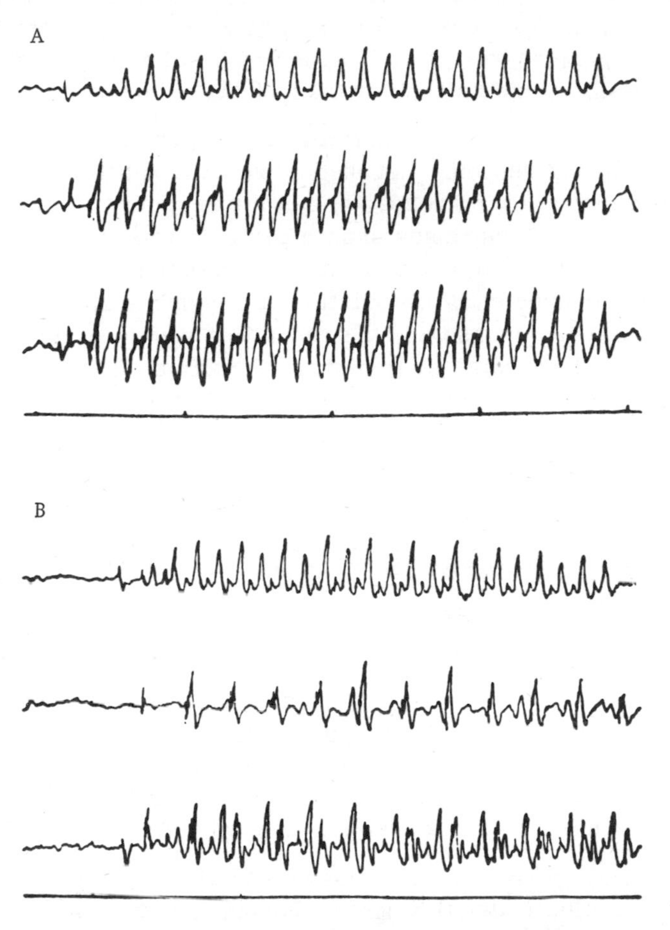

Fig. 3. Reduction by callosal stimulation of the response evoked by contralateral cortical stimulation; corpus callosum previously sectioned. The corpus callosum was stimulated via a steel microelectrode inserted into its cut edge. In A and B, the traces from top to bottom are, respectively: the responses evoked by contralateral cortical stimulation, the responses evoked by callosal stimulation, and the responses for combined stimulation. In the experiment shown in A, the cortical stimulating and recording electrodes were placed on the visual area, and the frequencies of stimulation of the cortex and the corpus callosum were the same; when both stimuli were presented, basically only the response evoked by callosal stimulation appeared. In the experiment shown in B, the cortical stimulating and recording electrodes were placed on the motor area, and the stimulus frequency for the corpus callosum was half that for the cortex. For the cortical and callosal stimuli that occur simultaneously, it is mainly the callosal response that appears, but for the cortical stimuli that occur alone, the very large, mainly surface-negative, responses reappear. Time markers: 1 sec.

thalamus; a distinct electrical response followed each stimulus, the amplitude being very large in the vicinity of the medial part, such as the dorsomedial nucleus. Although the responses were dissimilar, their amplitudes were relatively stable, and no periodic waxing and waning was present. Thus, the response was already apparent for the very first stimulus, and the amplitude of the former was the same as that for each of the subsequent stimuli.

From the general behavior of the waxing-and-waning response, it is apparent that it is very similar to the "recruiting response" evoked by repetitive stimulation of the medial portion of the thalamus [6-12] in the following respects: (1) the response is mainly surface-negative; (2) the amplitudes of successive responses show a periodic waxing and waning; (3) for stimulation at approximately 8/sec, the frequency of the responses and that of the stimuli are similar, but for relatively higher frequencies of stimulation, the responses do not readily appear. Evidently the waxing-and-waning response and the recruiting response differ only in that the latter is evoked by stimulation of the medial portion of the thalamus directly, whereas the former is evoked from the medial portion of the thalamus which is in turn activated by stimulation of the contralateral cortex. In other words, the waxing-and-waning response is a result of nerve impulses evoked by stimulation of the contralateral cortex transmitted through the medial portion of the thalamus to evoke activity in the region of the ipsilateral dorsomedial nucleus.

The Problem of the Nonappearance of the Response in the Absence of Callosal Sectioning

From comparison of Fig. 1A and Fig. 1B, we can note the point that prior to sectioning of the corpus callosum, repetitive stimulation evokes only a callosal response, but not the waxing-and-waning response. In some experiments, although the amplitude of the negative potentials of the waxing-and-waning response obtained after sectioning of the corpus callosum was 2 or 3 times greater than that of the negative potentials of the callosal response prior to callosal sectioning, the waxing-and-waning response was nonetheless not seen prior to callosal sectioning.

Further, on an animal whose corpus callosum had already been sectioned, we inserted another pair of steel microelectrodes through the cortex into the cut end of the ipsilateral corpus callosum. If the corpus callosum were stimulated via this pair of electrodes while the waxing-and-waning response was present, then the amplitude of the latter was greatly reduced, even to the point of disappearing. Figure 3 shows two examples of the results. In Fig. 3A and Fig. 3B from top to bottom, waxing-and-waning responses, callosal responses, and responses resulting from combined stimulation of the contralateral cortex and the corpus callosum, respectively, are shown. In Fig. 3A, the frequency of stimulation of the contralateral cortex and that of the corpus callosum are the same; in Fig. 3B, the frequency of stimulation of the corpus callosum is half of that for stimulation of the contralateral cortex. It is apparent that under the former conditions, when the two types of stimuli act simultaneously, each response of the waxing-and-waning response is greatly reduced. Under the latter conditions, in the absence of callosal stimulation, the amplitude of the waxing-and-waning response is very large as before, but during callosal stimulation, its amplitude is greatly reduced.

If the same frequency of stimulation were used for the contralateral cortex and for the corpus callosum, but the time relationship of the two stimuli were changed, it was apparent that during each callosal response, within the latency interval for the appearance of the responses evoked by stimulation of the contralateral cortex, the response evoked by the latter stimulus clearly decreased, even to the point of disappearing. Because the amplitude of the waxing-and-waning response was not very stable, we did not measure the time course of the effect of callosal stimulation on the form of the waxing-and-waning response. On the other hand, irrespective of the relation between the time intervals for the appearance of these two types of responses, the callosal response could not be significantly decreased as a result of stimulation of the contralateral cortex.

The above results clearly indicate that the waxing-and-waning response can be greatly diminished, even to the point of disappearing, by nerve impulses transmitted via the corpus callosum.

DISCUSSION

In the results reported in this paper, a point worthy of note is that in the absence of sectioning of the corpus callosum, the waxing-and-waning response cannot be evoked by contralateral cortical stimulation. It can also be mentioned that afferent nerve impulses from the corpus callosum can not only cause the waxing-and-waning response to diminish or disappear, but can also result in the disappearance of cortical responses evoked by direct thalamic stimulation of the dorsomedial nucleus. These results have two possible explanations: (1) The neuronal structures participating in the formation of the waxing-and-waning response are activated by afferent impulses from the corpus callosum, and when callosal responses are present, these neuronal structures become inexcitable, so that the waxing-and-waning response does not appear; (2) afferent impulses from the corpus callosum have an inhibitory effect on the waxing-and-waning response. From the following facts, it is apparent that the second explanation is not very likely: when each response of the waxing-and-waning response appears before the callosal response, the former does not have a prominent effect on the latter. It is also apparent that, following sectioning of the corpus callosum, stimulation of the cut end of the corpus callosum not only can result in a decrease or a disappearance of the waxing-and-waning response, but can also result in a diminution or disappearance in the cortical response evoked from the dorsomedial nucleus by stimulation of the contralateral cortex. This result not only indicates that activity from the corpus callosum has an inhibitory effect on the waxing-and-waning response, but also indicates that the site of the effect is not limited to the cortex but also includes the medial part of the thalamus, having in fact a major effect on the medial part of the thalamus. The pathways for the production of this effect are until now unclear, but two possibilities can be considered: (1) afferent impulses from the corpus callosum activate certain cells in the cortex having an inhibitory effect on the medial part of the thalamus, (2) in the corpus callosum there are fibers that course directly to the medial part of the thalamus. It should be mentioned that histological data on whether there are fibers coursing directly to the subcortical sites is so far lacking, although in view of the fact that callosal fibers do have collaterals [13], this possibility still exists. Ajmone Marsan and Morillo [14] reported that impulses passing from the corpus callosum to cells in the lateral geniculate body have an inhibitory effect; although they did not carry out further analyses to determine whether there were inhibitory fibers in the corpus callosum which passed directly to the lateral geniculate body, or whether impulses from the corpus callosum first activated certain cells of the cerebral cortex which had an inhibitory effect on the lateral geniculate.

The periodic alteration in the amplitude of the responses evoked by repetitive cortical stimulation is evident not only in the waxing-and-waning response but also in other responses, such as the response evoked by direct cortical stimulation. Clare and Bishop [15] attempted to explain these types of phenomena in terms of an alteration in the recovery cycle of the apical dendrites of pyramidal cells. In our experiments, it was observed many times that in the callosal responses evoked by repetitive stimulation, there was also a periodic change in the amplitude of the surface-negative potential that followed the initial surface-positive potential. However, we observed that after removal of the thalamus by suction, the periodic change in amplitude of this negative potential was no longer apparent. It is apparent that whether this phenomenon appears or not is dependent on whether the thalamus is present. It is quite possible that the thalamus has a periodic effect on the excitability of the cerebral cortex, and consequently potentials conducted to the cortex exhibit a periodic change of amplitude. Moreover, in animals under comparatively light barbiturate anesthesia, the periodic alteration of the amplitude of this response appears to be relatively small or almost inapparent. Considering the fact that such doses of anesthetic have a powerful effect on the brain stem reticular formation [16], their adverse effect on the periodic alteration of amplitude of the responses is also of interest in relation to the above-mentioned view.

There already exists no small amount of data (for example, [7, 17]) concerning the important role of the medial part of the thalamus in the connections between the two hemispheres. The results reported in the present paper indicate that the connections between the two hemispheres that are effected via the medial portion of the thalamus and via the corpus callosum interact with one another in a complex manner. These results also present new data concerning the view (see, for example, [18-20]) that the medial portion of the thalamus participates in the modulation of "spontaneous" electrical activity of the cerebral cortex.

SUMMARY

After sectioning of the corpus callosum, with direct stimulation of the cerebral cortex with low-frequency repetitive stimuli, after several stimuli, each stimulus evokes a primarily surface-negative potential from the contralateral cortex, the amplitude of which increases and decreases periodically. If the thalamus is then sectioned along the midline, this response disappears. At the same time that the responses are present at the cortex, a clear response following each stimulus can also be recorded from an electrode inserted into the medial portion of the thalamus on the side showing the cortical response. Afferent impulses from the corpus callosum have an inhibitory effect on this response.

REFERENCES

1. Fan Shi-fang and Shen Ke-fei. [Cortical repetitive responses elicited by a single contralateral stimulus] (in Chinese), Acta Physiol. Sinica, 25:114-118 (1962). (English translation on pp. 10-15 of this volume.)
2. Sawyer, C. H., Everett, J. W., and Green, J. D., The rabbit diencephalon in stereotaxic coordinates, J. comp. Neurol., 101:801-824 (1954).
3. Marshall, W. H., An application of the frozen sectioning technique for cutting serial sections through the brain, Stain Tech., 15:133-138 (1940).
4. Curtis, H. J., Intercortical connection of corpus callosum as indicated by evoked potentials, J. Neurophysiol., 3:407-413 (1940).
5. Chang, H.-T., Cortical response to activity of callosal neurons, J. Neurophysiol., 16:117-131 (1953).
6. Dempsey, E. W., and Morison, R. S., The production of rhythmically recurrent cortical potentials after localized thalamic stimulation, Am. J. Physiol., 135:293-300 (1942).
7. Jasper, H., Diffuse projection systems: The integrative action of the thalamic reticular system, Electroenceph. clin. Neurophysiol., 1:405-420 (1949).
8. McLardy, T., Diffuse thalamic projection to cortex: an anatomical critique, Electroenceph. clin. Neurophysiol., 3:183-196 (1951).
9. Hanbery, J., and Jasper, H., Independence of diffuse thalamocortical projection system shown by specific nuclear destruction, J. Neurophysiol., 16:252-273 (1953).
10. Verzeano, M., Lindsley, D. B., and Magoun, H. W., Nature of the recruiting response, J. Neurophysiol., 16:183-195 (1953).
11. Jasper, H., Functional projection of the thalamo-reticular system, in: Brain Mechanisms and Consciousness, C. C. Thomas (1954), pp. 374-401.
12. Kerr, F. W. L., and O'Leary, J. L., The thalamic source of cortical recruiting in the rodent, Electroenceph. clin. Neurophysiol., 9:461-476 (1957).
13. Ramón y Cajal, S., Histologie du Systeme Nerveux de l'Homme et des Vértebrés, Vol. II, (1955), pp. 578-579.

14. Ajmone-Marsan, C., and Morillo, A., Cortical control and callosal mechanisms in the visual system of the cat, Electroenceph. clin. Neurophysiol., 13:553-563 (1961).
15. Clare, M. H., and Bishop, G. H., Potential wave mechanisms in cat cortex, Electroenceph. clin. Neurophysiol., **8**:583-602 (1956).
16. French, J. D., Verzeano, M., and Magoun, H. W., A neural basis of the anesthetic state, Arch. Neurol. Psychiat., 69:519-529 (1953).
17. Mei Chen-tung, Liu Jen-yi, Sung Li-fen, and Yu Hui-chung [Studies on interocular transfer of conditioned reflex in cats] (in Chinese), Acta Physiol. Sinica, 25(3):191-197 (1962).
18. Morison, R. S., and Dempsey, E. W., A study of thalamo-cortical relation, Am. J. Physiol., 135:281-292 (1942).
19. Ralston, B., and Ajmone-Marsan, C., Thalamic control of certain normal and abnormal cortical rhythms, Electroenceph. clin. Neurophysiol., 8:559-582 (1956).
20. Shen Ke-fei and Fan Shi-fang [Repetitive surface-negative cortical responses evoked by single stimuli] (in Chinese), Acta Physiol. Sinica, 22:167-174 (1958).

CORTICAL EXCITABILITY CHANGES FOLLOWING TRANSCALLOSAL AFFERENT EXCITATION*

Zhang Gin-ru

Department of Physiology, First Medical College of Shanghai, Shanghai

In cats anesthetized with sodium pentobarbital, it has been found that stimulation of the corpus callosum or the contralateral homotopic point has an inhibitory effect on potentials evoked by tone bursts in the auditory area of the cortex [1]. In results obtained from the "encéphale isolé" preparation, however, there was additionally a facilitatory effect [2]. The conditions under which these inhibitory and facilitatory phenomena appear await clarification. The present investigation on rabbits constitutes a further analysis of this question, and also examines the question of effects of drugs, etc. This paper will report the results of these investigations.

METHODS

The experiments were carried out with 51 rabbits, using as the conditioning response the callosal potential evoked by stimulation, with a bipolar electrode, of the homotopic point of the sensorimotor area of the contralateral cerebral cortex, and using as the test response the potential evoked by stimulation of the superficial radial nerve. The methods of stimulation and recording were the same as those reported in a previous paper [3].

RESULTS

1. Temporal Aspects of the Excitability Changes Following Transcallosal Afferent Excitation

Following the callosal potentials (conditioning response) evoked by stimulation of the homotopic point of the contralateral cortex, the amplitude of the potential evoked by stimulation of the superficial radial nerve (the test response) shows an initial inhibitory and a later facilitatory phenomenon. In the first trace of each of the groups in Fig. 1A, it can be seen that the test responses show a complete or partial inhibition; in each of the third and fourth traces, a clear facilitatory phenomenon can be seen. In particular there is an especially distinct increase of the negative wave of the test response. Under conditions of comparatively lighter anesthesia, the conditioning response often has an afteractivity; consequently, on a background of this facilitation, accompanying the afteractivity there may be a corresponding oscillation of

*Acta Physiologica Sinica, 26:321-327 (1963).

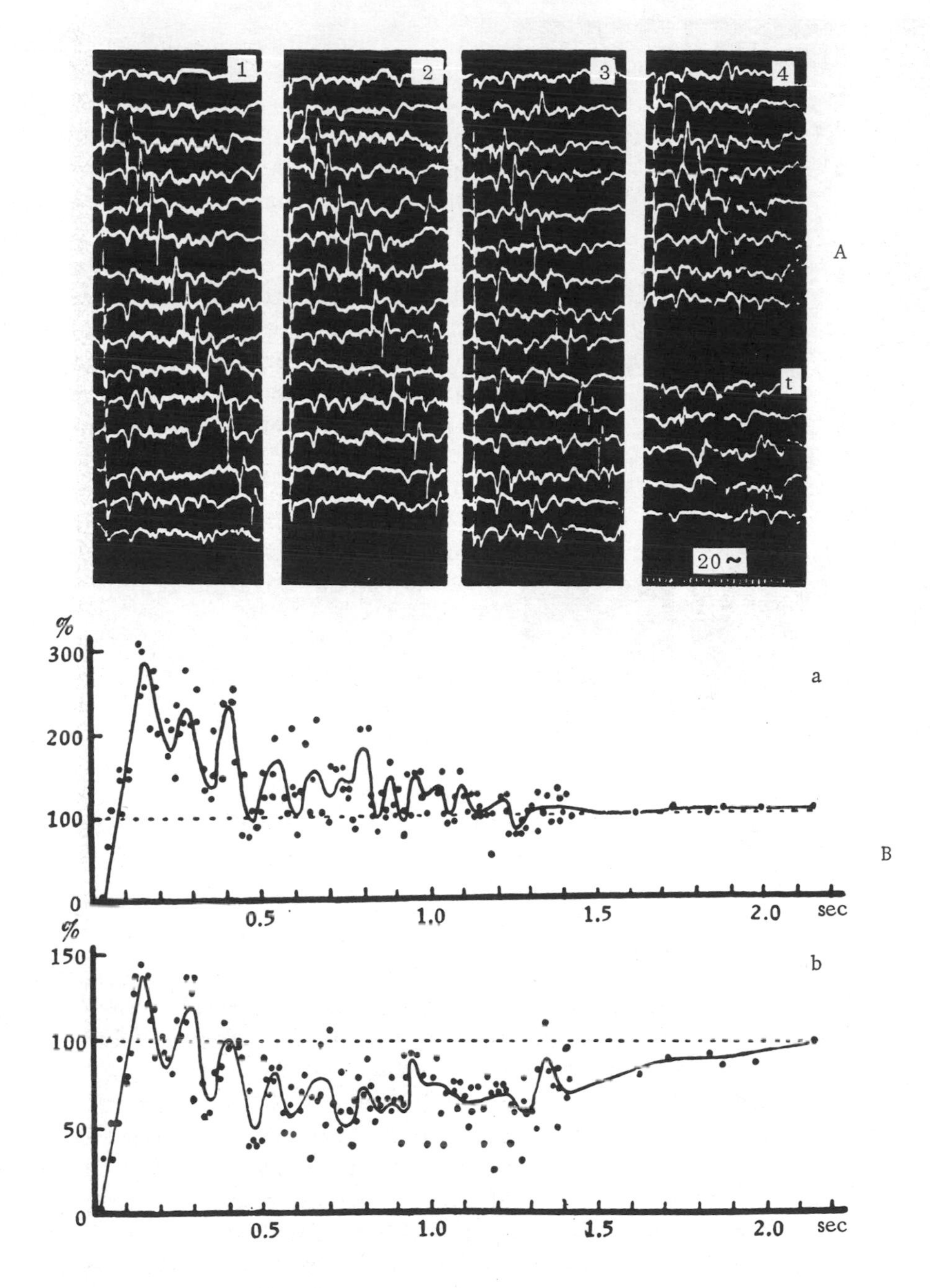

Fig. 1. Time factor of the excitability change following stimulation of the homotopic point on the contralateral cerebral cortex. A. Recordings of potentials, t indicates the control recordings of evoked potentials from stimulation of the superficial radial nerve (test response). In the recordings in Groups 1, 2, 3, and 4, the first response is the callosal potential resulting from stimulation of the homotopic point of the contralateral cortex (conditioning response), and the following one is the sequence of the test response, led from an electrode in the middle of the region yielding the test response. B. Curve of excitability change, of which part of the results are drawn from A: a) results using the amplitude of the negative wave of the test response as a measuring index; b) results employing the amplitude of the positive wave of the test response as a measuring index. Ordinate: measured value relative to control value in percentage; abscissa: time interval between conditioning and test stimuli.

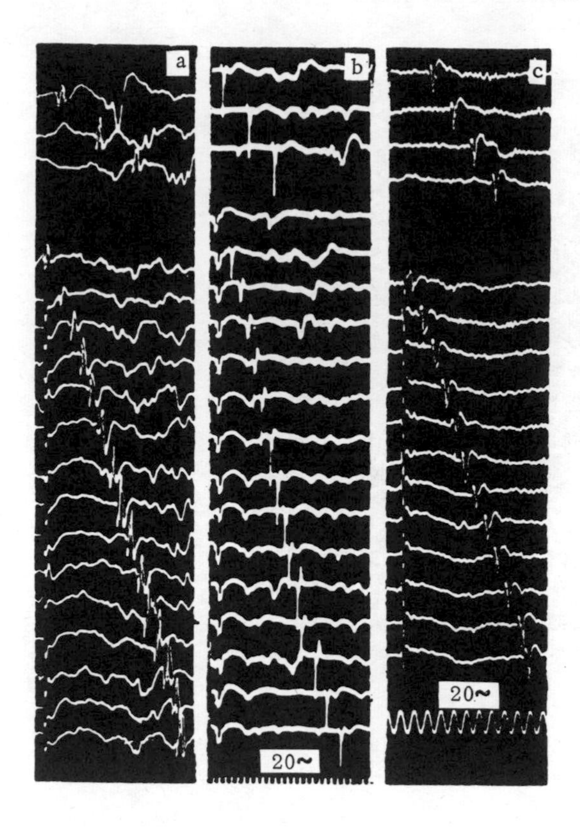

Fig. 2. Results under different experimental conditions. In each group, the upper traces are the controls for the test responses; of the lower records, the first is the conditioning response, the following ones are the sequence of the test response. a) Evoked potential from the edge of the area giving responses to electrical stimulation of the superficial radial nerve (test response), conditioning stimulation delivered to homotopic point of the contralateral cortex. b) Conditioning and test stimuli presented separately to the left and to the right superficial radial nerves, recording electrode in center of test-response area of the left cortex. c) "Encéphale isolé," recording electrode in center of area giving evoked potential to stimulation of the inferior orbital nerve, conditioning stimulation delivered to homotopic point of the contralateral cortex.

the excitability. During the course of the positive waves of the aftereffect, the facilitation is increased (as in the third traces of Group 1, Fig. 1A); during the course of the recovery, the facilitation is diminished (as in trace 3 of Group 3 in Fig. 1A). In the afterperiod of the conditioning response, the positive waves of the test response may exhibit a secondary inhibitory phenomenon (as in the 12th through 14th traces of Group 2 in Fig. 1A). Figure 1B is the corresponding experimentally obtained curve of the excitability changes. Curve a is the result using the amplitude of the negative wave of the test response as an index. From this curve, it can be seen that the initial inhibitory period was approximately 100 msec. There is an absolute and a relative refractory period; the later facilitation is maximum at approximately 150 msec after the stimulation, and can exceed 200%. The entire facilitatory period extends out to approximately 1.2 sec. Curve b is the result using the amplitude of the positive wave of the test response as an index of measurement; the initial absolute and relative refractory period also extends out to approximately 100 msec, but the absolute refractory period itself is comparatively short. The later facilitation does not exceed 150%, and persists for only 200-300 msec, approximately. Following this facilitation, the positive waves manifest a second inhibitory course; the inhibition can exceed 30-40% and can persist to the vicinity of 1.5 sec. In these two curves, a correspondence of the oscillations of the excitability can be seen.

There are some differences between our results and those from the "encéphale isolé" preparation [2]. In our results (a) the facilitation persists longer, and (b) a secondary inhibition of the positive waves appears following the facilitation. This difference could have been due to the anesthesia; consequently we carried out an investigation using rabbit "encéphale isolé," the test response being changed to the potential evoked by stimulation of the inferior orbital nerve. The results were obtained under conditions of light anesthesia. The excitability changes following stimulation of the homotopic point of the cortex included, as before, an initial inhibition and a subsequent facilitation. Moreover, the positive waves of the test response had, as before, a secondary inhibitory period, but the complete time course was clearly shorter than those mentioned above (Fig. 2C). The time course of this facilitation and the results from the cat are very similar. From this it can be seen that the effect of anesthesia on cortical excitability is

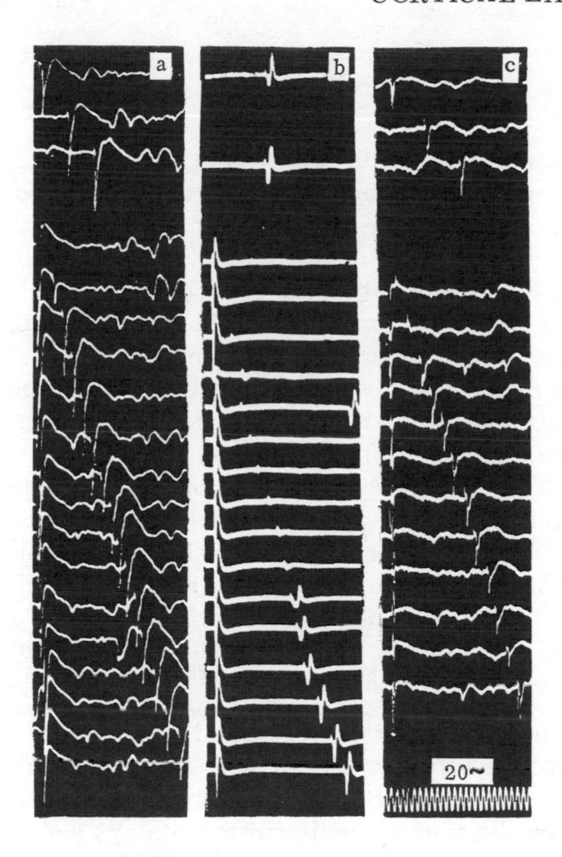

Fig. 3. Effects of different drugs. The top traces in each group are the control evoked potentials from stimulation of the superficial radial nerve (test response). Of the lower traces, the first is the callosal potential following stimulation of the homotopic point of the contralateral hemisphere, the subsequent sequence is that of the test response. a) Following use of 3% gamma-aminobutyric acid; b) following 3% strychnine; c) following 2% procaine. Amplification factor in b is $^1/_3$ of that in a and c.

principally a prolongation of its time course; this conclusion is in agreement with that of a previous paper [3].

Under conditions of weak conditioning stimuli, there is no absolute refractory period in the initial inhibition and the following facilitatory phenomenon is also comparatively small. If the intensity of the conditioning stimulus is increased, an absolute refractory period does appear, and the later facilitation is also clearly increased. If excessively strong conditioning stimuli are used, the initial inhibitory period is prolonged, and the subsequent facilitatory phenomenon is reduced.

2. Spatial Aspects of Excitability Changes Following Transcallosal Afferent Stimulation

If the recording is made from the edge of the region giving an evoked response to electrical stimulation of the superficial radial nerve (test response), the positive waves of the response proper become smaller, but the negative waves in contrast clearly are larger. At this time, if the callosal potentials resulting from stimulation of the contralateral homotopic point are used as conditioning response, there is an initial inhibition and a subsequent facilitation. During the facilitatory period, the increase in the amplitude of the negative wave of the test response is especially clear (Fig. 2A).

In another group of experiments, the recording electrode was in the middle of the region giving the test response, and if the conditioning stimulating electrode were gradually moved from the homotopic point, then the inhibitory and facilitatory phenomena both gradually decreased, if the conditioning stimuli used were not excessively strong.

3. Results of Conditioning Stimulation Applied to Peripheral Nerves

The above conditioning responses were from stimulation of the homotopic point of the contralateral cortex, and as a consequence such a conditioning response necessarily includes an antidromic callosal component. To insure that the conditioning response did not involve this component, in one group of experiments, the conditioning and the test stimuli were separately presented to the left and right superficial radial nerves. If the cutaneous branch of the superficial radial nerve projecting to the contralateral cortex is stimulated while the recording electrode is on the left sensorimotor region of the cortex, then the resulting potential change is altered significantly by the conditioning stimulation, via the callosally transmitted activity from the contralateral cortex. Under these conditions, the cortical excitability change has an initial inhibitory and a later facilitatory phase, and the positive waves of the test response show a secondary inhibition, as before (Fig. 2B). Following section of the fibers of the corpus callosum, the above-mentioned excitability changes do not appear. Consequently, under natural conditions, passage of activity of the callosum to the opposite side definitely possesses an early inhibitory and a later facilitatory effect. This conclusion is the same as that of earlier workers [2].

4. Effect of Sodium Pentobarbital

In one group of experiments, we selected sodium pentobarbital, instead of a chloralose – urethane mixture, for anesthesia, in order to examine the effect of the former drug. In the results, it was seen that sodium pentobarbital employed in a dose of approximately 40 milligrams per gram of body weight, had no clear effect on the facilitation; the callosal facilitatory phenomenon was clearly seen as before. The facilitatory phenomenon disappeared only for very deep anesthesia, such that the animal's respiratory activity had stopped, necessitating artificial respiration. The inhibitory phenomenon remained, however.

5. Effects of Different Drugs Applied to the Surface of the Cortex

Using 2-mm-square pieces of filter paper soaked with 3% gamma-aminobutyric acid which were applied to the surface of the cortex in such a way that the potential changes could be led from the center of the filter paper, the early appearance of the callosal facilitatory phenomenon was rapidly lost, but both the initial and the secondary inhibition of the positive waves remained intact (Fig. 3A). After the gamma-aminobutyric acid had been washed off with physiological saline, the callosal facilitatory effect once again reappeared.

The results obtained with 2% procaine were the same as those for gamma-aminobutyric acid (Fig. 3C). From this finding, it is evident that both of the above-mentioned drugs clearly impede the callosal facilitation, but do not change its inhibitory effect.

When 3% strychnine was applied, the conditioning and the test responses were clearly increased, the time interval of the primary inhibition was greatly prolonged, and the facilitatory phenomenon was lost (Fig. 3B). At this time, the cortex showed spontaneous "strychnine spikes." Following each "strychnine spike" there was an inhibitory period, and if the control response proper fell within this inhibitory period, the test response appeared disturbed immediately. For example, in the four traces in the lower part of Fig. 3B, the conditioning response evoked spontaneous "strychnine spikes" following the inhibitory period, which reduced the inhibitory effect on the test response.

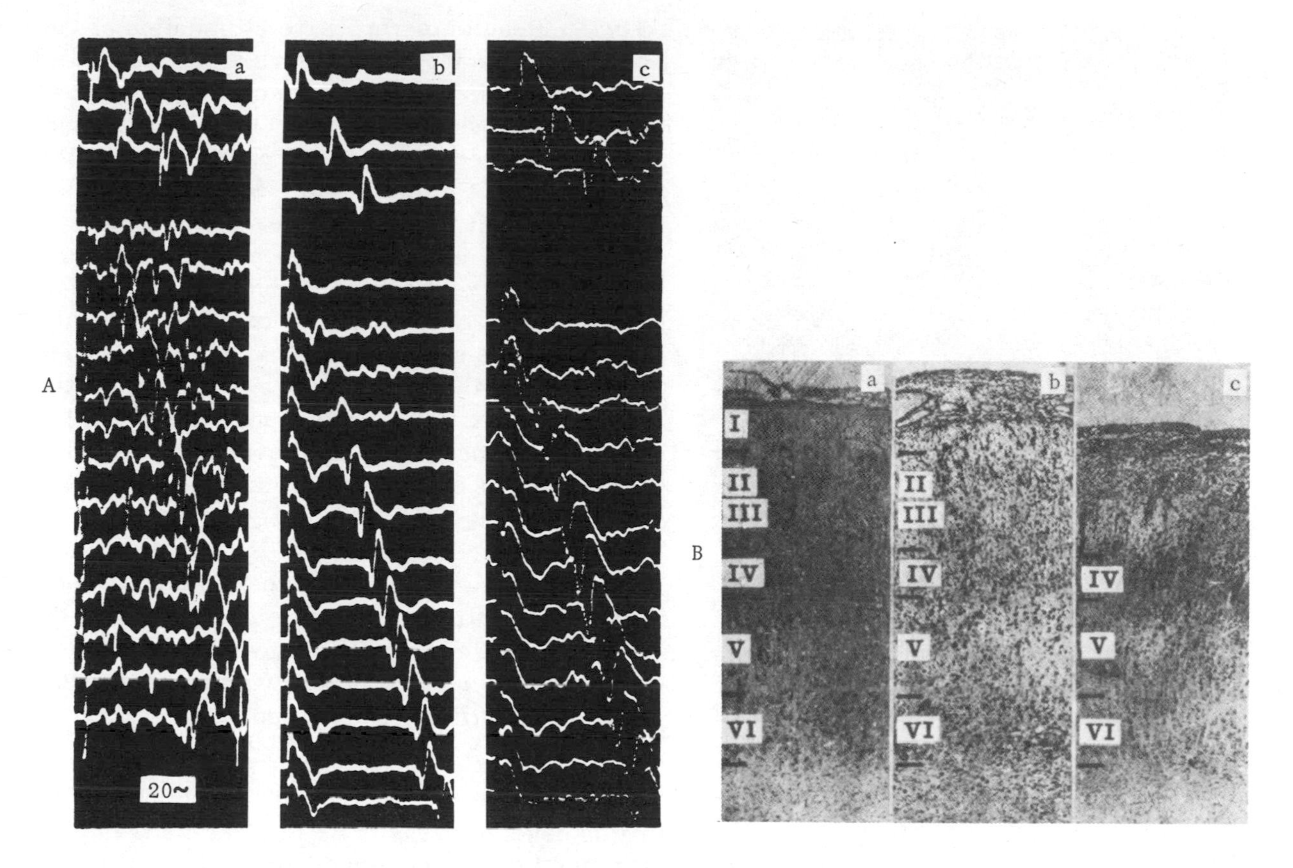

Fig. 4. Results following laminar thermocoagulation of the cortex. A. Recordings of potentials. The upper trace in each is the control trace for the potential evoked by stimulation of the superficial radial nerve (test response); of the lower traces, the first is the conditioning response resulting from stimulation of the homotopic point of the contralateral cortex, the following one is the sequence of the test response. B. Photographs of sections of cortical structure: a) control without the application of thermocoagulation; b) following application of thermocoagulation at 80°C for 1 sec to the surface of the dura mater; c) following application at 80°C for 2 sec. a and b are results from the cortex of the two hemispheres of the same animal; c is the result for another animal.

6. Laminar Analysis of the Callosal Facilitation

In order to determine which particular cortical layer was most closely related to the callosal facilitation, we employed 15-micron (inside-diameter) glass microelectrodes filled with 0.9 saturation NaCl, and recorded the potentials as the tip of the electrode was gradually advanced through the cortical layers. In this way it could be ascertained whether there were any changes in the respective locations of the positive and negative waves of the test response, prior to and following the facilitation. The results showed that the position of the waves before and after the facilitation did not change, both positive waves became negative upon penetrating the cortex beyond 0.7 mm, but the negative wave became positive at 0.9 mm. Consequently, the structures of no specific layer could be related to the callosal facilitation.

In other experiments, we used the laminar thermocoagulation method of Dusser de Barenne [4], in order to probe the localization of the callosal facilitation. The animal was first operated upon, and heat of 80°C was applied to the dura mater by means of a 5-mm-diameter chromium-plated electric iron, which resulted in an increase in the temperature of the cortex for one or two seconds. The experiments were carried out 5 to 6 days after the operation. Following

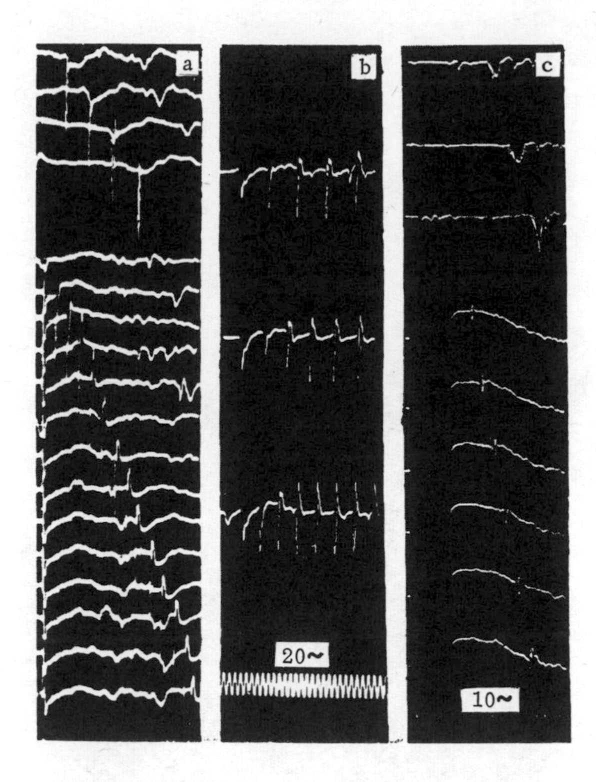

Fig. 5. Results of repetitive stimulation of the homologous point of the contralateral cortex. a) Conditioning and test stimuli both applied to the homologous point of the contralateral cortex; the upper traces are control photographs of the test response; of the lower traces, the first is the conditioning response, the subsequent ones are the sequence of the test response. b) Results of low-frequency repetitive stimulation of the homologous point of the opposite cortex. c) Results for repetitive stimulation at a somewhat higher frequency of the homologous point of the contralateral cortex. The upper traces are control photographs of the potential evoked by stimulation of the superficial radial nerve (test response, peaks of positive waves retouched). Of the lower traces, the first of the series of responses is the conditioning response from repetitive stimulation of the homologous point of the contralateral cortex, the following are the sequence of the test responses. Overall time constant of amplifier in c changed to 0.9 sec.

thermocoagulation by 80°C for one second, there was death of nerve cells in layer I and in part of layer II (Fig. 4B: b), but 80°C for 2 seconds resulted in death of nerve cells in layers I, II, and III (Fig. 4B: c). Under the former conditions, the facilitatory capability remained intact (Fig. 4B: b), but under the latter conditions, the facilitation was completely lost (Fig. 4A: c). It was thus apparent that the facilitation was mainly a manifestation of activity in structures of layers II and III. The initial inhibition also remained intact following the application of the above-mentioned thermocoagulation.

7. Results of Repetitive Stimulation of the Contralateral Homotopic Point

Previous workers [5, 6], using paired stimuli of specific intervals delivered to the contralateral homotopic point, reported that a facilitatory effect on a second callosal response can be seen following an early callosal response. We attempted to examine whether this facilitation were not the same as the facilitation resulting from potentials evoked by contralaterally-transmitted somesthetic stimuli. Experiments showed that if suitable conditioning and test stimuli were both presented to the contralateral homologous point, then a prior callosal response indeed had a facilitatory effect on the negative wave of a subsequent response (Fig. 5a). From the time course, it was evident that this facilitation was the same as the facilitation of evoked responses from contralateral somesthetic transmission.

When we used a suitable frequency of repetitive stimulation, inducing a repetitive callosal response, it was seen that the negative wave of each response had a tendency of gradually increasing (Fig. 5c). This phenomenon is similar to that of the appearance of the slow negative component upon direct repetitive stimulation of the cortex [7]. At the time of the appearance of the slow negative potential, the cortical excitability is clearly decreased, and the potential evoked from the superficial radial nerve is clearly inhibited (Fig. 5c). If the frequency of the repetitive stimulation is still further increased, then the slow negative potential change becomes still greater, the decrease in cortical excitability also becomes clearer, and the facilitatory phenomenon is lost.

DISCUSSION

The present experimental work on the rabbit further establishes the existence of the callosal facilitation. Sodium pentobarbital does not have a clear effect on this facilitation, a result that is not in agreement with the results of previous workers [2]. We consider that the reason for not observing a clear facilitation with sodium pentobarbital anesthesia could be (1) the intensity of the conditioning stimulus may not have been sufficiently great; (2) the time course examined was comparatively short. Moreover, in this investigation, it was also seen that the positive waves of the test response had a secondary inhibitory period, but previous workers did not report this point [1, 2]; a possible reason is that we used the sensorimotor region of the cortex as the object of investigation, in contrast to the experiments of previous workers who used the auditory area of the cortex of cats. At the same time, the time course of secondary inhibition of the positive waves required a comparatively longer time course of observation in order to be seen.

Some workers have considered that gamma-aminobutyric acid can selectively block so-called "excitatory synapses" [8], and that strychnine can block so-called "inhibitory synapses" [9], but these two substances can both result in a loss of the callosal facilitation. At the same time, procaine has an anesthetic effect on nerve fibers and likewise can cause the callosal facilitation to be lost. Consequently, it would seem that the capacity of callosal facilitation could not be explained in terms of changes in the activity of so-called "excitatory" or "inhibitory synapses."

The site in the cortex of termination of callosal projections is presently still disputed. Important terminations both in the superficial cortex (layers I, II, and III) [10] and in the deep layers (layers III-VI) [11] are recognized. From the results of the present investigation, it is apparent that callosal facilitation is principally the result of activity of structures in the second and third layers, which then gives rise to the facilitatory effect of fiber projections terminating in the superficial layers. Application of thermocoagulation to the superficial layers, however, did not eliminate the initial inhibitory effect, which indicates that the callosal projection also has terminations in the deep layers.

SUMMARY

Employing as conditioning responses callosal potentials evoked by stimulation of the homotopic point of the contralateral cerebral hemisphere and as test responses the potentials evoked by stimulation of the superficial radial nerve, we found that the cortical excitability consisted of an early inhibitory and a later facilitatory change. The facilitatory phenomenon was manifested principally by changes of the negative wave of the test response, and following the facilitation of the positive waves, there was also a secondary inhibitory period. If the conditioning response had an afterdischarge, then the cortical excitability change was also accompanied by a corresponding periodic oscillation. The local application of gamma-aminobutyric acid, procaine, or strychnine, could all result in loss of the callosal facilitatory phenomenon, but sodium pentobarbital anesthesia had no great effect on it. Following death of cortical layers II and III by application of thermocoagulation, the callosal facilitatory phenomenon was lost. Repetitive stimulation of the homologous point of the contralateral cortex can give rise to a slow negative potential change, at which time the cortical excitability is clearly diminished.

The assistance of Su Zhao-jiang, in the preparation of the sections, and of Xu Ning-shan in the participation in this work, are gratefully acknowledged.

REFERENCES

1. Chang, H.-T., Interaction of evoked cortical potentials, J. Neurophysiol., 16:133-144 (1953).
2. Bremer, F., Physiology of the corpus callosum, Res. Publ. Ass. nerv. Dis., 36:424-448 (1958).
3. Zhang Gin-Ru [Interaction of evoked cortical potentials in the rabbit] (in Chinese), Acta Physiol. Sinica, 26:165-171 (1963). (English translation in this volume, pp. 1-9.
4. Dusser de Barenne, J. G., Laminar destruction of the nerve cells of the cerebral cortex, Science, 77:546-547 (1933).
5. Peacock, S. M., Activity of anterior suprasylvian gyrus in response to transcallosal afferent volleys, J. Neurophysiol., 20:140-155 (1957).
6. Grafstein, B., Organization of callosal connections in suprasylvian gyrus of cat, J. Neurophysiol., 22:504-515 (1959).
7. Caspers, H., and Baedecker, W., Shifts and cortical DC potentials produced by local serial stimuli and their importance for the production of convulsions, Electroenceph. clin. Neurophysiol., 12:259 (1960).
8. Purpura, D. P., Girado, M., and Grundfest, H., Selective blockade of excitatory synapses in the cat brain by gamma-aminobutyric acid (GABA), Science, 125:1200-1201 (1957).
9. Purpura, D. P., and Grundfest, H., Physiological and pharmacological consequences of different synaptic organizations in cerebral and cerebellar cortex of cat, J. Neurophysiol., 20:494-522 (1957).
10. Chang, H.-T., Cortical response to activity of callosal neurons, J. Neurophysiol., 16:117-131 (1953).
11. Nauta, W. J. H., Terminal distribution of some afferent fiber systems in the cerebral cortex, Anat. Rec., 118:333 (1954).

THE INTERACTION OF CALLOSAL POTENTIALS AND POTENTIALS EVOKED BY THALAMIC STIMULATION*

Zhang Gin-ru

Department of Physiology, First Medical College of Shanghai, Shanghai

In the cat, previous workers have reported that there is a mutually inhibitory effect between cortical potentials evoked by stimulation of the thalamic somatosensory nucleus, and callosal potentials [1]. Subsequently, it was seen that the latter, in addition, has a facilitatory relationship to the former [2, 3]. In the present work, the author has observed, in rabbits, that the former also has a facilitatory effect on the latter, and the mutually facilitatory effect of both was investigated. This paper reports these results.

METHODS

Experiments were carried out with the domestic rabbit, with the aid of a stereotaxic atlas [4], using a pair of electrodes with an interelectrode distance of 2 to 3 mm for stimulation of the nucleus ventralis posterior of the thalamus, the resulting cortical evoked potential constituting the conditioning response. The response evoked by stimulation, with an electrode pair, of the homologous point on the contralateral cerebral cortex was the test response. The methods of stimulation, recording, etc., were the same as that reported previously [5].

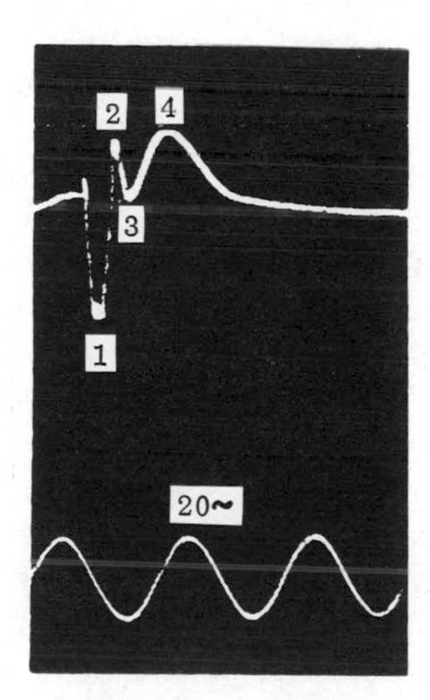

Fig. 1. Individual wave components of the callosal potential.

RESULTS

1. Time Course of the Effect on Callosal Potentials of Cortical Potentials Evoked By Stimulation of the Nucleus Ventralis Posterior of the Thalamus

The typical callosal potential is composed of four waves (Fig. 1). The courses of waves 1 and 2 are comparatively fast; they are termed the fast component. The courses of waves 3 and 4 are comparatively slow, and they are called the slow component. The amplitudes of the fast and slow components of the callosal poten-

*Acta Physiologica Sinica, 27:348-355 (1964).

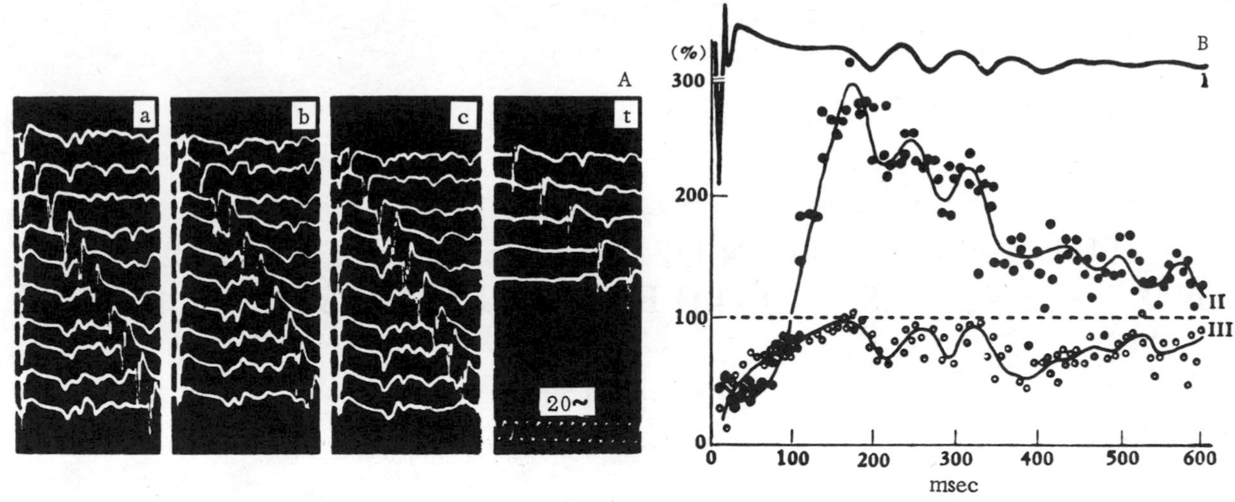

Fig. 2. Time course of the effect on callosal potentials of cortical potentials evoked by stimulation of the nucleus ventralis posterior of the thalamus. A. Recordings of potentials. t shows the control recordings of the callosal potentials (test response). a, b, and c are recordings of the effect at different periods of time; the first response in each trace is the potential evoked by stimulation of the n. ventralis posterior (the conditioning response), the subsequent responses are the successive fluctuations of the test responses. B. Curves of excitability change, obtained from A. Ordinate: percentage of amplitude of test response. Abscissa: interval (in msec) between conditioning and test stimuli. Curve II is the result using the amplitude of the slow component of the test response (peak-to-peak value of waves 3 and 4) as the index of measurement. Curve III is the result using the amplitude of the fast component (peak value of wave 1) as the index. Curve I is the potential change of the conditioning response.

tials (the test response) are clearly altered after the potentials evoked by stimulation of the n. ventralis posterior (the conditioning response) appear. For example, in the first and second traces of Fig. 2A: b, the amplitudes of the fast and slow components of the test response are both decreased; however, in other traces of b the fast component remains comparatively small, whereas the slow component is clearly larger. Accompanying the appearance of the afterdischarge of the conditioning response, the amplitude of the slow component also manifests periodic waves: during the development of the positive waves of the afterdischarge, the increase in amplitude is especially clear (see c, trace 3, and a, trace 4), but during the course of the recovery of the positive wave, the increase in amplitude is comparatively unclear (see b, trace 4, and c, trace 4). Figure 2B indicates the curves of the excitability effects obtained from this group of experiments; curve II is the result using the amplitude of the slow component (peak-to-peak value of waves 3 and 4) as an index. From this curve, one can see an early inhibitory and a later facilitatory, excitability change. The inhibition extends out to approximately 100 msec, whereas the facilitation, which is maximum at approximately 180 msec after stimulation, and can reach approximately 300%, persists beyond 0.5 sec. Curve III is the result using the amplitude of the fast component (peak value of wave 1) as an index; from this curve only an inhibition is seen. Curve I is the potential change of the conditioning response, and from comparison of curves I, II, and III, it can be seen that the excitability is increased during the course of development of the positive waves of the afterdischarge of the conditioning response, but during the recovery phase, the excitability is decreased somewhat.

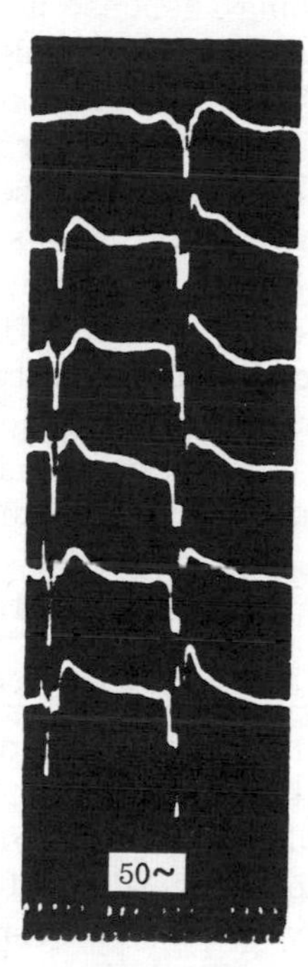

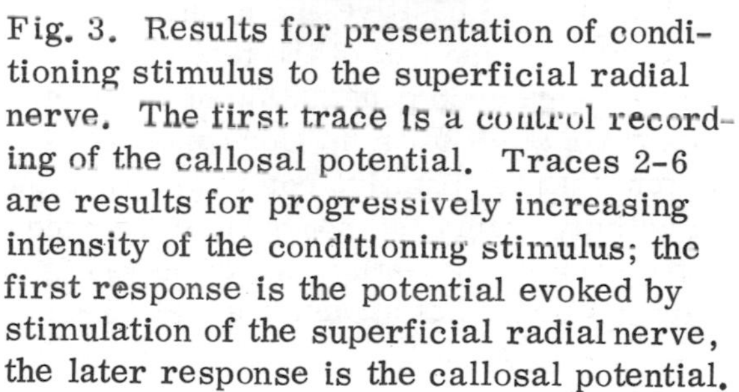

Fig. 3. Results for presentation of conditioning stimulus to the superficial radial nerve. The first trace is a control recording of the callosal potential. Traces 2-6 are results for progressively increasing intensity of the conditioning stimulus; the first response is the potential evoked by stimulation of the superficial radial nerve, the later response is the callosal potential.

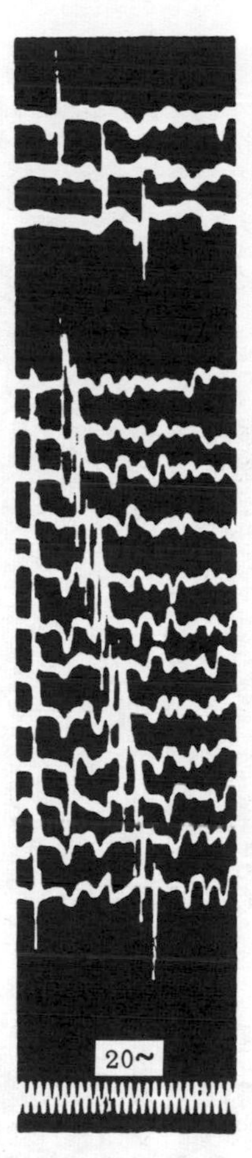

Fig. 4. Reinforcement of the afterdischarge of the callosal potentials by cortical potentials evoked by stimulation of the n. ventralis posterior. The top three traces are control recordings of the callosal potentials. In each of the lower traces, the first response is the cortical evoked potential from stimulation of the n. ventralis posterior. Following it are the sequence of the callosal potentials.

In order for the facilitation of the amplitude of the slow component of the test response to appear, both the conditioning and the test stimuli must be of sufficient intensity. Increasing the intensity of the conditioning stimulation increases the degree of facilitation, and extends its duration.

If the conditioning stimulation is delivered to the superficial radial nerve, the results obtained are the same as the above-mentioned ones; the facilitatory phenomenon also appears only if there is an increased amplitude of the slow component of the test response, and if the intensity of the conditioning stimulus is increased, the increase of amplitude of the slow component also becomes clearer (Fig. 3).

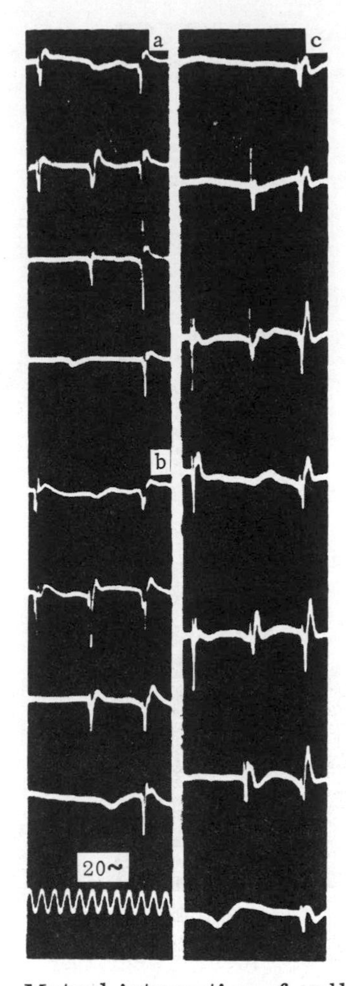

Fig. 5. Mutual interaction of callosal potentials and cortical potentials evoked by stimulation of the n. ventralis posterior. a and b are experimental results from the same animal, those in c are for another animal. For detailed comment, see text.

2. Reinforcement of the Afterdischarge of Callosal Potentials by Cortical Potentials Evoked by Stimulation of the Nucleus Ventralis Posterior

The conditioning response not only resulted in a clear increase in the amplitude of the slow component of a following test response, but also clearly reinforces its afterdischarge. In the control recordings of the test response in Fig. 4, the afterdischarge is not at all obvious, but under the effect of the conditioning response, the afterdischarge is quite evident. The time interval between this reinforced afterdischarge activity and the primary component of the test response is constant, and hence it cannot represent the afterdischarge of the conditioning response.

3. Results of Laminar Glass Microelectrode Recordings

By using 15-micron (inside-diameter) glass microelectrodes filled with 32% (0.9 saturation) sodium chloride solution, and recording the electrical potentials upon stepwise advancement through cortical layers, it was seen that the abovementioned facilitated positive and negative potentials of the slow component of the test response both reversed sign upon penetration of the cortex beyond approximately 0.7 mm. Consequently it was apparent that the positive potentials were the result of activity from deep neuronal structures, and that the negative potentials were the result of activity from superficial structures.

4. Interaction of Callosal Potentials and Cortical Potentials Evoked by Stimulation of the Nucleus Ventralis Posterior

Callosal potentials can have a facilitatory effect on potentials evoked by stimulation of the somesthetic nucleus of the thalamus [2, 3], and previous workers have termed this callosal facilitation. In the present work, the author found that the latter also has a facilitatory effect on the former, from which it could be supposed that the callosal potentials, after being increased by this facilitation, could result in a reinforced callosal facilitation. In order to test this deduction, we employed single evoked responses from stimulation of the nucleus ventralis posterior (test response) as controls (Fig. 5a, trace 4, and 5c, traces 1 and 7). Then, paired stimuli were used — callosal potentials evoked by conditioning stimuli, and test stimuli the same as mentioned above — and the callosal facilitatory phenomenon was apparent (Fig. 5a, trace 3, and 5c,

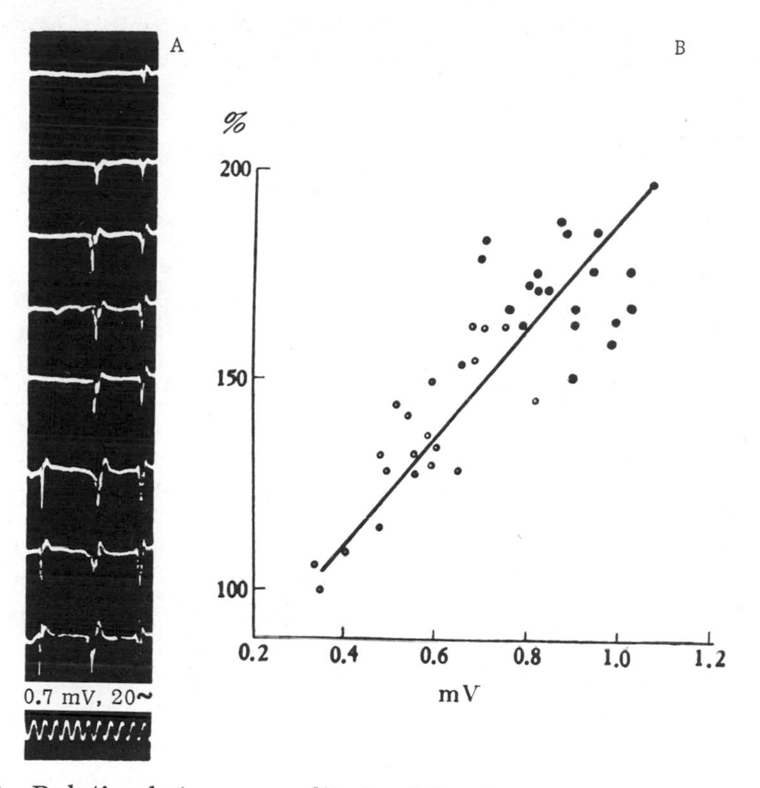

Fig. 6. Relation between amplitude of the slow component of the callosal potential and callosal facilitation. A. Recordings of the potentials; for explanation see text. B. Graph of relation between amplitude of the slow component and the facilitation. Abscissa: amplitude of the slow component (peak-to-peak value of waves 3 and 4). Ordinate: percentage of amplitude (peak-to-peak value) of potential evoked by stimulation of the n. ventralis posterior. Open circles are results of the effect for simple callosal potentials; filled circles are results in which the callosal potentials are preceded by the thalamic-evoked potentials.

trace 2). Afterwards, upon using three stimuli, the first delivered to the n. ventralis posterior, and the second and third being the same as the above-mentioned conditioning and test stimuli, respectively, it was apparent that as the slow component of the callosal potential increased, the amplitude of the test response clearly increased (Fig. 5a, trace 2, and 5c, trace 3). When this intermediary callosal potential was eliminated, then the first evoked potential did not have such an enhancing effect on the following evoked potential (Fig. 5a, trace 1, and 5c, trace 4), from which it is apparent that the intermediary callosal potential really does have an important effect, consequently verifying the above-mentioned deduction. It should be pointed out, however, that when the intermediary was changed to the potential evoked by stimulation of the n. ventralis posterior, the result was similar (Fig. 5c, trace 5), because under the condition that the successive two evoked responses are separated by an appropriate time interval, the former can have a facilitatory effect on the latter (Fig. 5c, trace 6).

Likewise, it could be supposed that a potential evoked by thalamic stimulation, which had been intensified by the facilitatory effect of callosal potentials, could in turn have an enhanced facilitatory effect on the slow component of a following callosal potential. In order to test this supposition, we used callosal potentials (test responses) resulting from single stimuli as a control

(Fig. 5b, trace 4). Then, paired stimuli were used, the conditioning evoked potentials resulting from stimulation of the n. ventralis posterior, and the test stimuli being the same as the above; an increase in the slow component of the callosal potential could be seen (Fig. 5b, trace 3). Next, three stimuli were used, the first evoking the callosal potential, the second and third being the same as for the above-mentioned conditioning and test responses, respectively; it was apparent that following the increase in amplitude of the potential evoked by stimulation of the n. ventralis posterior, the slow component of the test response clearly increased (Fig. 5b, trace 2). If this intermediary evoked potential were omitted, then the first callosal potential had no such intensifying effect on the following one (Fig. 5b, trace 1), which indicates that this intermediary evoked potential definitely has an important effect.

5. Relation between the Slow Component of the Callosal Potentials and the Callosal Facilitatory Effect

From the above results, it is apparent that an increase in the amplitude of the slow component of callosal potentials increases the callosal facilitation. Consequently, in one series of experiments, we employed varying-intensity stimulation of the homologous point on the contralateral hemisphere, the resulting callosal potentials being preceded by potentials evoked by stimulation of the n. ventralis posterior, which caused the slow component of the callosal potentials to be of different amplitudes. The changes in the callosal facilitation were then examined in an attempt to find the relationship between the two. The first trace of Fig. 6A is a control recording of an evoked potential from stimulation of the n. ventralis posterior. If this evoked potential is preceded by the slow component of callosal potentials of different amplitudes, then the callosal facilitatory phenomenon can be seen; as the amplitude of the slow component increases, its facilitatory effect also increases (Fig. 6A, traces 2-5). If the callosal response is preceded by a potential evoked by stimulation of the n. ventralis posterior, then as the amplitude of its slow component increases, the facilitatory effect also progressively increases (Fig. 6A, traces 6-8). The amplitude of the slow component of the callosal potentials, and the amplitude (peak-to-peak value) of the potentials evoked by stimulation of the n. ventralis posterior were measured, as shown in Fig. 6B; it is apparent the two are directly proportional. The open circles are the results of the effect for simple callosal potentials, whereas the filled circles are the results for the conditions under which the callosal potentials were preceded by the evoked potentials. The open circles are all distributed in the left lower part of the graph, whereas the filled circles are all in the right upper portion, which indicates that only under the conditions that the callosal potentials are preceded by the evoked potentials is there a clear increase in its slow component, and a clear reinforcement of the callosal facilitation.

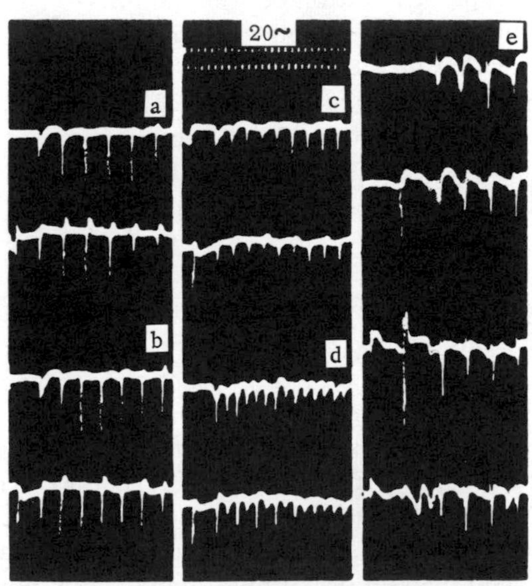

Fig. 7. Form of cortical responses evoked by repetitive stimulation of the n. ventralis posterior during increased cortical excitability. a, b, c, and d are experimental results for the same animal; e shows the results for another animal. For detailed explanation see text.

6. Alteration of Form of Cortical Responses Evoked by Repetitive Stimulation of the Nucleus Ventralis Posterior during Increased Cortical Excitability

By use of a suitable frequency of repetitive stimulation of the n. ventralis posterior, the cortex can be induced to produce a series of evoked potentials, which at first gradually increase, the third or fourth frequently being the largest, the following ones gradually decreasing again (Fig. 7a, b, c, e, first traces). This phenomenon is the same as that observed by previous workers in examining the responses of the auditory area of the cortex upon repetitive stimulation of the medial geniculate body [6]. When this series of evoked potentials was preceded by a callosal potential, the cortical excitability was increased, so that the form of the response of the series of evoked potentials changed. The initial responses were clearly increased, the second response frequently being the largest, the following ones gradually becoming smaller (Fig. 7a, b, c, e, second traces). At the same time, for a somewhat higher frequency of stimulation, the responses had the form of successive alternation of amplitudes (Fig. 7b, c, second traces). Upon still higher frequency of repetitive stimulation, the series of evoked potentials no longer had an initial increasing and a subsequent decreasing sequence; instead, there was a gradual diminution of the responses right from the beginning, and the succession of alternation appeared immediately (Fig. 7d, trace 1). On this occasion, a preceding callosal potential causes this alternation of responses to become more clearly evident (Fig. 7d, trace 2). The form of the alternating size of the responses could quite possibly arise from a subsequent response falling in the nonresponsive interval of the recovery cycle of the preceding response; consequently, increasing the frequency of stimulation facilitates the appearance of this form of response; at the same time, during increased cortical excitability the response is increased, but the unresponsive portion of the recovery cycle is prolonged, which favors the appearance of the successive alternation.

When single stimuli were delivered to the n. ventralis posterior preceding a callosal potential, causing the slow component of the callosal potential to be increased, and the cortical excitability was increased to the extreme, then of the series of evoked potentials arising from repetitive stimulation of the n. ventralis posterior, the first response became the largest, the subsequent ones gradually becoming smaller (Fig. 7e, trace 3). When this intermediary callosal response was omitted, the above-mentioned result did not appear (Fig. 7e, trace 4). This phenomenon is completely in agreement with the increased cortical excitability seen in the cat under conditions of constant illumination of the retina [7].

DISCUSSION

From the results of the present work, it can be seen that callosal responses and cortical responses evoked by stimulation of the thalamic somatosensory nucleus have a mutually facilitatory effect. With respect to one side of the cortex, there is not simply an increase in excitability resulting from contralaterally transmitted callosal responses, but also an increase due to somesthetically transmitted activity, which by increasing the callosal response transmitted from the contralateral side, results in a progressive increase in excitability. From this it can be seen that the mutual facilitation between callosal potentials and evoked potentials arising from stimulation of the thalamic somatosensory nucleus can further reinforce cortical activity. This point is of considerable significance with respect to the coordination of activity between homologous cortical points.

In a previous paper [3], it was mentioned that the callosal facilitatory effect and the activity of superficial cortical structures are related; in the present work it can be seen that this facilitation is directly proportional to the amplitude of the slow component (principally the negative potential, see Fig. 6A) of the callosal potential. From results of microelectrode recordings from different cortical layers, it was apparent that the negative potential of the slow component is derived from superficial cortical structures; from this it is additionally clear that the callosal facilitation is unquestionably related to the activity of superficial cortical structures.

Concerning the relationship between the afterdischarge and the cortical excitatory changes, previously mentioned by earlier workers [8] and in previous papers [3, 5], the same conclusion as before was obtained in the present work, callosal potentials being used as test responses. It is apparent that the afterdischarge activity, irrespective of whether it originates from direct stimulation of the cortex, from somesthetically-transmitted activity, or from callosally-transmitted activity, has the same result, namely, during the course of development of the positive waves of the afterdischarge, cortical excitability is increased, but during its recovery phase, cortical excitability is diminished.

SUMMARY

Using cortical potentials evoked in the rabbit by stimulation of the nucleus ventralis posterior of the thalamus as conditioning responses, and callosal potentials as test responses, it was seen that the latter manifested an initial inhibitory, and a subsequent facilitatory, change. The course of the facilitation was manifested as an increase in the amplitude of the slow component of the test response and a reinforcement of its afterdischarge, whereas the fast component showed only an inhibition. If the conditioning response had an afterdischarge, then the amplitude of the test response would also be correspondingly periodic. If the callosal potentials, being increased by the effect of potentials evoked from stimulation of the n. ventralis posterior, in turn act upon potentials evoked by stimulation of the n. ventralis posterior, then a powerful callosal facilitatory effect can be seen; the larger the amplitude of the slow component of the callosal potential, the greater is its callosal facilitatory effect, the two being directly proportional.

REFERENCES

1. Chang, H.-T., Interaction of evoked cortical potentials, J. Neurophysiol., 16:133-144 (1953).
2. Bremer, F., Physiology of the corpus callosum, Res. Publ. Ass. nerv. ment. Dis., 36: 424-448 (1958).
3. Zhang Gin-ru [Cortical excitability changes following transcallosal afferent excitation] (in Chinese), Acta Physiol. Sinica, 26:321-327 (1963). (English translation appears in this volume, on pp. 24-32.)
4. Sawyer, C. H., Everett, J. W., and Green, J. D., The rabbit diencephalon in stereotaxic coordinates, J. comp. Neurol., 101:801-824 (1954).
5. Zhang Gin-ru [Interaction of evoked cortical potentials in the rabbit] (in Chinese), Acta Physiol. Sinica, 26:165-171 (1963). (English translation appears in this volume on pp. 1-9.)
6. Hanbery, J., and Jasper, H., Independence of diffuse thalamocortical projection system shown by specific nuclear destruction, J. Neurophysiol., 16:252-271 (1953).
7. Chang, H.-T., and Pressman, Y. M., Effect of monocular illumination on cortical response to optic nerve stimulation, Scientia Sinica, 11:1249-1258 (1962).
8. Chang, H.-T., Changes in excitability of the cerebral cortex following a single electric shock applied to the cortical surface, J. Neurophysiol., 14:95-111 (1951).

THE EFFECT OF ELECTRIC STIMULATION OF THE BRAIN STEM ON THE GALVANIC SKIN REFLEX*

Li Peng, Cheng Jie-shi, and Sun Zhong-han

Department of Physiology, First Medical College of Shanghai, Shanghai

Wang [1-7] demonstrated that the midbrain reticular formation has a facilitatory effect and that the reticular formation of the medulla has an inhibitory effect, respectively, on the galvanic skin reflex (GSR), and he concluded that the specific locations in the reticular formation that facilitate or inhibit the galvanic skin reflex were perhaps not identical with the centers found by Magoun et al. to be related to facilitation and inhibition of somatic reflexes. In the present work, the effect on galvanic skin activity of electrical stimuli applied systematically to the brainstem reticular formation, point-by-point with the aid of a stereotaxic apparatus, is examined.

METHODS

The experiments were carried out on 36 cats weighing between 2 and 4 kg, anesthetized with urethane (700 mg/kg) and chloralose (35 mg/kg) administered intraperitoneally. A cannula was inserted into the trachea, the common carotid artery was ligated bilaterally, and the cranial cavity was opened at different points on the calvarium. When it was necessary to stimulate the pons or the medulla, the cerebellar tentorium was first removed and the occipital part of the cerebrum was removed by suction, for an advantageous exposure; the cerebellum was not usually removed, however. After the operation, bone wax, powder, and sponges were used to stop bleeding, following which the head of the cat was affixed in the stereotaxic apparatus. One of the channels of a spray-type ink-writing instrument (Mingograf [8]), was used for recording the blood pressure of the right common carotid artery; another channel recording the palmar galvanic skin activity. Respiration, recorded by means of a Marey tambour from the motion of the thoracic cavity, or as pressure within the trachea, was recorded above the curves of blood pressure and galvanic skin activity. The GSR was evoked by electrical stimulation of the right tibial nerve, the stimuli being pulses of amplitude 8 to 51 volts, i.e., of sufficient amplitude to obtain a clear GSR; the frequency was usually 47/sec, the pulse width 0.5 msec, and the duration of the stimulation was 5 sec. For stimulating the brain stem centers, bipolar electrodes were used [9] which consisted of a pair of glass-insulated nickel wires or an insulated stainless steel tube containing a varnished wire, the tips of the electrodes being bare for 0.2 m, the outside diameter being 0.5 mm, and the interelectrode distance being 0.2 to 0.3 mm. These stimuli were also pulses, of an intensity of 4 to 8 V, a frequency of 51/sec, a duration of 0.5

*Acta Physiologica Sinica, 29(1):26-33 (1966).

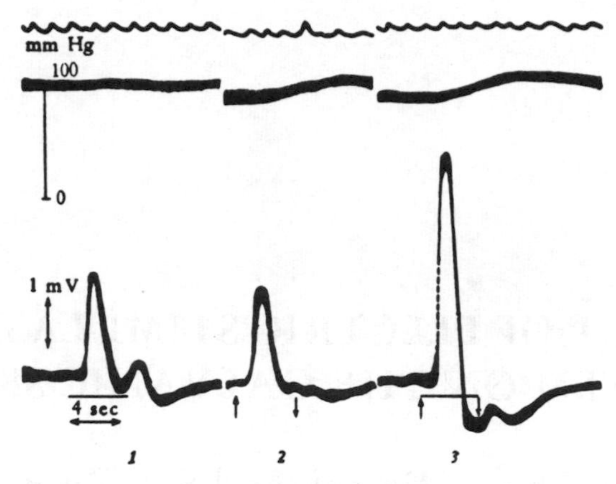

Fig. 1. First type of effect of brain stem stimulation on galvanic skin activity. Cat, 3 kg. 1) Stimulation at P5, R3, H-8, stimulus intensity 4 V (stimulus frequency 51/sec, pulse duration 0.5 msec, duration of stimulation 5 sec; the same for the following figures), evokes clear galvanic skin activity. 2) Stimulation of the proximal portion of the right tibial nerve, stimulus intensity 51 V (stimulus frequency 57/sec, pulse width 0.5 msec, stimulus duration 5 sec; the same for the following figures); there is a clear galvanic skin reflex. 3) Simultaneous stimulation of the brain stem and the proximal portion of the right tibial nerve; there is a facilitation of the galvanic skin reflex. Upper curve in each figure: movement of the thoracic cavity (an upward deflection indicates inspiration); middle curve: blood pressure in the right common carotid artery; lower curve: galvanic skin activity of the left anterior palm (subcutaneous potential difference between the volar and dorsal skin of the left forepaw (an upward deflection indicates a drop of skin potential on the volar side). Symbols: ———, stimulation of brain stem; ↑↓, stimulation of the right proximal portion of the right tibial nerve (the same for the following figures).

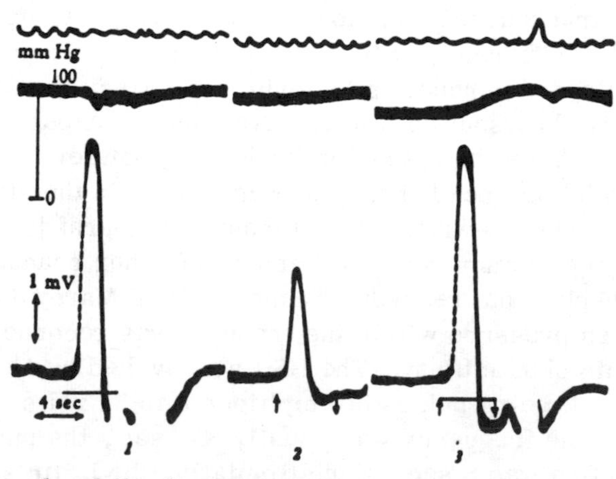

Fig. 2. Another example of the first type of effect of brain stem stimulation on galvanic skin activity. Animal the same as for Fig. 1. 1) Stimulation of brain stem (electrode position the same as in Fig. 1, stimulus strength 8 V) evokes prominent galvanic skin activity. 2) Stimulation of the proximal portion of the right tibial nerve (stimulus intensity 51 V); there is a clear galvanic skin reflex. 3) Simultaneous stimulation of the brain stem and the proximal portion of the right tibial nerve; masking of the galvanic skin reflex is apparent.

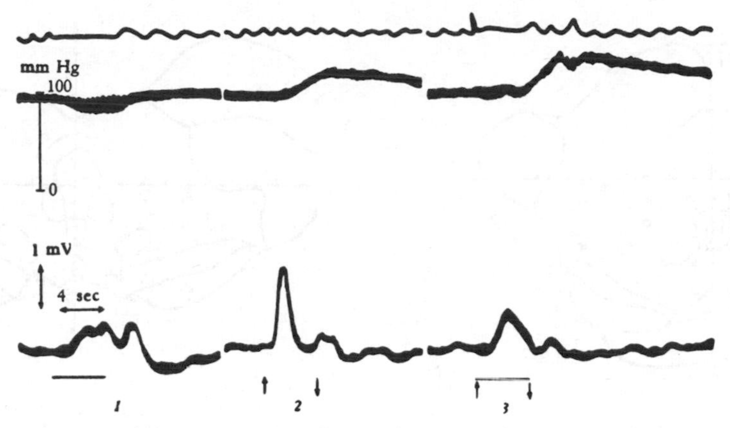

Fig. 3. Second type of effect of brain stem stimulation on galvanic skin activity. Cat, 3 kg. 1) Stimulation of brain stem (P8, R5, H-6; stimulus intensity 8 V); the evoked galvanic skin activity is small and develops slowly; following cessation of stimulation there is a rebound. 2) Stimulation of the proximal portion of the right tibial nerve (stimulus intensity 51 V); there is a clear galvanic skin reflex. 3) Simultaneous stimulation of the brain stem and the proximal portion of the right tibial nerve; the figure shows an inhibition of the galvanic skin reflex.

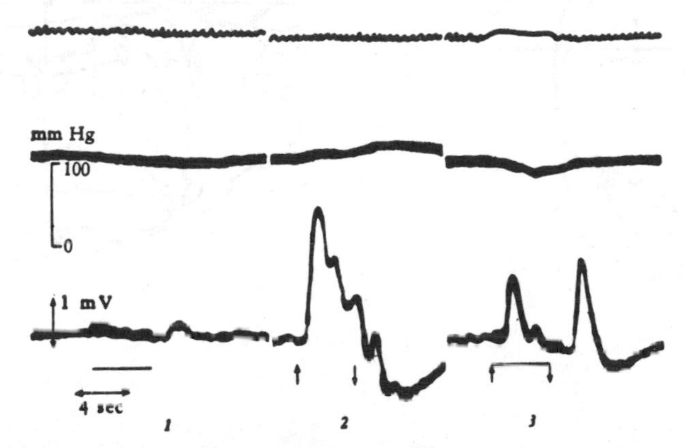

Fig. 4. Third type of effect of stimulation of the brain stem on galvanic skin activity. Cat, 2.25 kg. 1) Stimulation of the brain stem (P15, L2, H-10, stimulus intensity 8 V); no galvanic skin activity is evoked, but following termination of the stimulation there is a rebound. 2) Stimulation of the proximal portion of the right tibial nerve (stimulus intensity 51 V); there is a clear galvanic skin reflex. 3) Simultaneous stimulation of the brain stem and the proximal portion of the left tibial nerve; inhibition of the galvanic skin reflex is apparent and the rebound phenomenon is enhanced.

msec; the duration of each stimulation was 5 sec. The progression of the electrode placements was from surface to depth, from the midline to the left or right, and from anterior to posterior, for point-by-point stimulation at approximately 2-mm steps in the antero-posterior and left-to-right directions (the insertions being made at odd values for a part of the animals, and at even values for a part of the animals), the vertical steps being at intervals of 1 mm. After completion of an experiment, a 10% solution of formalin was injected into the distal end of the common carotid arteries, the head was removed, and after fixation for several days the brain stem was serially sectioned into 1-mm slices by means of a microtome, and the position of each stimulating electrode noted, thus verifying the location of each point of stimulation. The response for each point was then plotted on a figure.

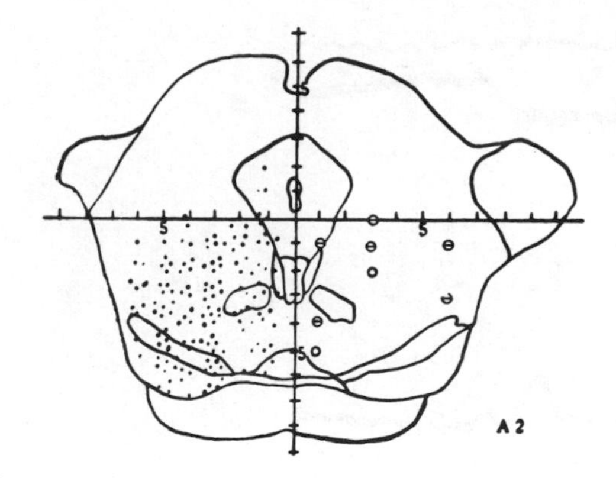
A 2

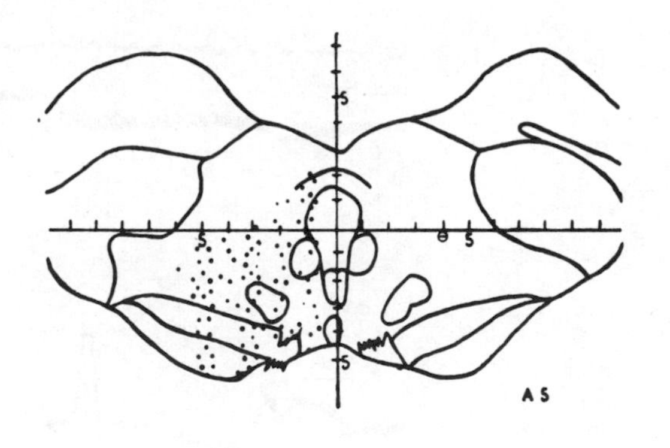
A 5

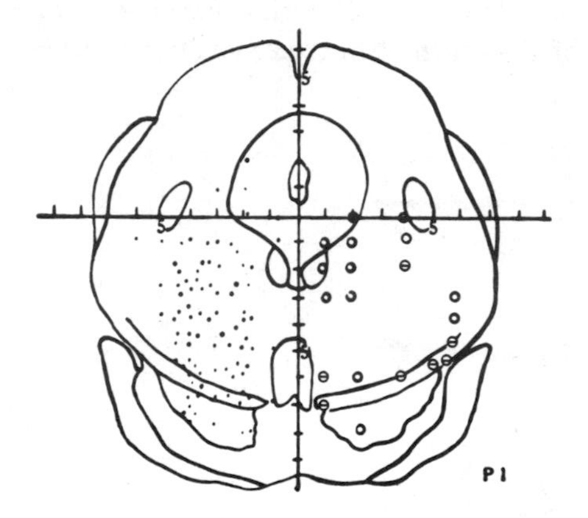
P 1

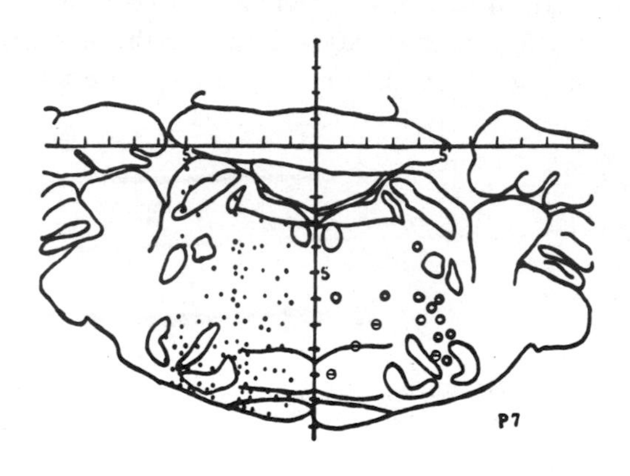
P 7

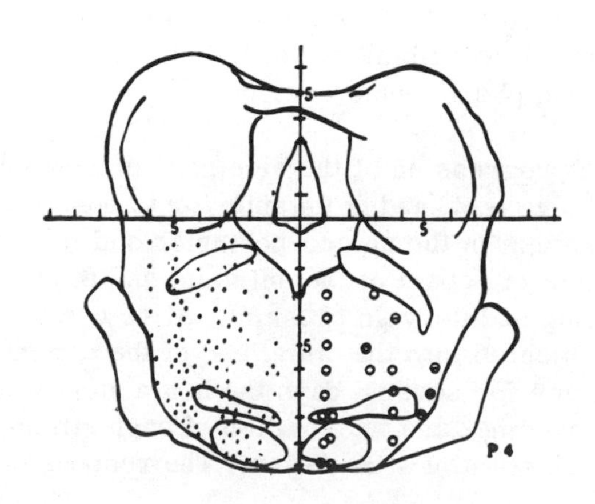
P 4

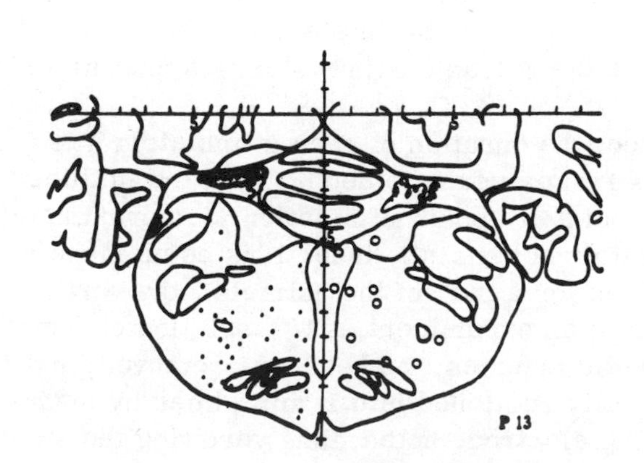
P 13

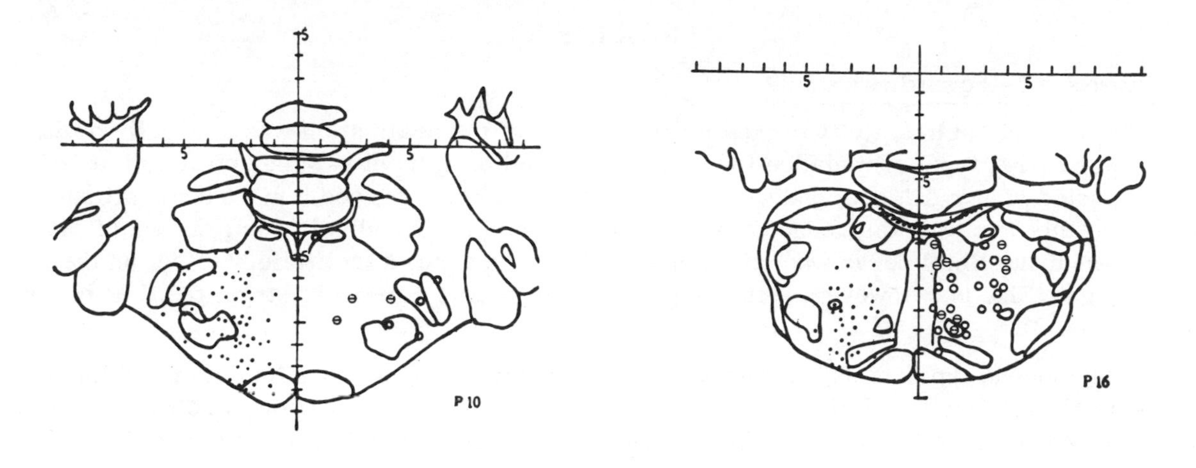

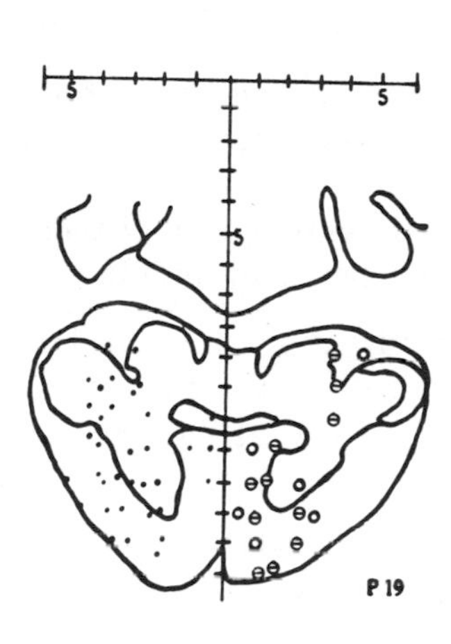

Fig. 5. Diagrams of sections of the brain stem of the cat showing the distribution of points facilitating and inhibiting galvanic skin activity, respectively. The figures are drawn from the experimental data for 36 cats; the meaning of the symbols for each section are as follows: ● points for the first type of response (see Fig. 1 or Fig. 2); ○ points for the second type of response (see Fig. 3); ⊖ points for the third type of response (see Fig. 4); A5, section through the anterior end of the superior colliculus and the anterior part of the midbrain; A2, section through the middle of the superior colliculus and the anterior part of the pons; P1, section through the posterior part of the superior colliculus and the middle of the pons; P4, section through the inferior colliculus and the posterior part of the pons; P7, section through the anterior part of the cerebellar peduncles and the trapezoid bodies; P10, section through the cerebellar peduncles and the cochlear nuclei; P13, section through the medial longitudinal fasciculus, the anterior vestibular nuclei, and the inferior olivary nuclei; P16, section 2 mm anterior to the obex; P19, section 1 mm posterior to the obex.

RESULTS

1. Types of Responses

The galvanic skin activity evoked by stimulation of the brain stem was generally of three types. In the first type, stimulation immediately evoked prominent skin potential activity with a latent period of 1 to 1.5 sec, and with an amplitude that could exceed 5 mV. If the stimulus strength at this location was reduced, the response was correspondingly smaller, but of the same type. If the tibial nerve was stimulated at the same time, a facilitatory effect on the GSR was apparent (Fig. 1). However, if the original response was relatively large, the GSR became masked (Fig. 2).

In the second type, no skin potential activity appeared during the stimulation, or the activity was small and developed slowly, but upon termination of the stimulation, a rebound was sometimes apparent; stimulation at this location simultaneously with stimulation of the proximal end portion of the tibial nerve inhibited the GSR (Fig. 3).

In the third type, there was no skin potential activity during stimulation, but upon cessation of the latter, there was a rebound; and stimulation at this location simultaneously with stimulation of the proximal portion of the tibial nerve inhibited the GSR and enhanced the rebound (Fig. 4).

In observations on GSR's evoked by stimulation of the proximal portion of peripheral nerves, Wang [10] also found a rebound response, and considered that this reflected the existence of simultaneous excitatory and inhibitory processes. The rebound phenomenon seen in the present experiments could indicate that stimulation of these points in the brain stem excited both facilitatory neurons and inhibitory neurons for the GSR.

In order to determine whether stimulus intensity, frequency, and duration affected the form of the response, we carried out observations at selected points; the results showed that increasing the stimulus frequency, intensity, or duration only increased the amplitude of the response but did not result in any change in its form.

The amplitude of the galvanic skin activity evoked in the four paws of the cats by stimulation on a particular side in the brain stem was sometimes the same, sometimes different, but the form of the response was uniformly the same.

2. Distribution of Facilitatory and Inhibitory Points

From the anterior border of the superior colliculus to 3 mm posterior to the obex, altogether 1732 points in the brain stem were stimulated. All of the results were plotted on a series of diagrams of brain-stem sections, each section being separated by 3 mm. Points that facilitated and inhibited galvanic skin activity were plotted on the left and right sides, respectively (Fig. 5). From the results it is apparent that points influencing galvanic skin activity are widely distributed in the entire brain-stem reticular formation. In a section passing through the superior collicus and the anterior end of the midbrain (see Fig. 5: A5), facilitatory points were predominant, inhibitory points appearing only rarely. At the level of the superior colliculus and the anterior end of the pons (see Fig. 5: A2), facilitatory points again predominated, although inhibitory points were comparatively numerous. As the sections pass more posteriorly through the level of the posterior part of the midbrain to the pons and the anterior border of the medulla (Fig. 5: P1, P4, P7, P10, P13 , it is apparent that facilitatory points gradually become fewer and inhibitory points increase. Posteriorly from 2 mm anterior to the obex (see Fig. 5: P16, P19), inhibitory points increase prominently, and facilitatory points decrease conspicuously. Thus, with respect to the brain stem as a whole, it is apparent that facilitatory points predominate in the midbrain reticular formation, but with gradual progression posteriorly, inhibitory points gradually increase, so that such points are mainly distributed in the reticular formation of the medulla. However, the distribution of facilitatory and inhibitory points is scattered and mutually interspersed without clearly delimited zones, although it is apparent that in general there are more facilitatory points than inhibitory ones.

3. The Relation between Galvanic Skin Activity and Respiratory and Blood-Pressure Responses

In the stimulation of 1169 points, we recorded respiratory and blood pressure simultaneously with galvanic skin activity. Among these, points whose stimulation altered galvanic skin activity were the most numerous, amounting to 1095 points, and were very widely distributed. The number of points capable of evoking a respiratory response numbered only 252; those which evoked a blood-pressure response totalled only 385. Among these a portion of the points at the same time evoked the previously-mentioned second or third types of behavior.

There were altogether 347 points that showed two kinds of responses simultaneously. Of these there were 179 points, or 52%, in which galvanic skin activity was accompanied by a blood-pressure response, of which the greater part were distributed in the midbrain. Among these, there was increase in galvanic skin activity and a rise in blood pressure for 140 points, and for 8 points an inhibition of galvanic skin activity was accompanied by a drop in blood pressure. Thus, there was a parallel response of galvanic skin activity and blood pressure for a total of 148 points, so that 83% of the points simultaneously showed the second type of behavior. There were altogether only 31 points, or 17%, for which the directions of change in the galvanic skin activity and the blood-pressure response were not the same. It is thus apparent that in the midbrain, although points that facilitate or inhibit galvanic skin activity are more widely distributed than the points that increase or decrease blood pressure, there were many points that increased blood pressure which were identical with, or adjacent to, facilitatory points, and there were a few inhibitory points that were coincident with those that reduced blood pressure.

There were altogether 141 points, or 41%, for which galvanic skin activity was accompanied by a respiratory response, most of which were in the caudal end of the midbrain, the pons, and in the medulla anterior to the obex. Galvanic skin facilitation was accompanied by an inspiratory response at 58 points, by expiration at 57 points, and by an acceleration of respiration at 7 points, whereas inhibition of galvanic skin activity was accompanied by expiration at 13 points, by inspiration at 5 points, and by acceleration of respiration at 1 point. It is apparent that points evoking a facilitation or inhibition of galvanic skin are generally not contiguous with the points that evoke expiratory or inspiratory responses.

There were an additional 27 points having simultaneous blood-pressure and respiratory responses, most of the points being in the medulla anterior to the obex. Among these the most, numbering 16 points, or approximately 60%, were those resulting in a drop of blood pressure accompanied by inspiration, from which it is evident that these are related to the hypotensive and inspiratory centers within the same region.

Points having responses in all three categories simultaneously totalled 162, most of which were in the pons; among these there were 54 points that resulted in a facilitation of galvanic skin activity accompanied by a rise in blood pressure and by inspiration or an acceleration of respiration, and there were 3 points that inhibited galvanic skin activity accompanied by a decrease in blood pressure and by expiration, from which it can be seen that the three types of responses were completely paralleled in altogether 57 points or approximately only 35%.

DISCUSSION

From these experiments it can be seen that points facilitating galvanic skin activity are completely predominant in the reticular substance of the cephalic end of the midbrain, but inhibitory points gradually increase more posteriorly, so that in the medulla, the proportion of inhibitory points is maximum, their quantity being almost equal to that of the facilitatory points. This finding confirms the conclusion of Wang that following transection of the brain stem between the superior and inferior colliculi, the disappearance of galvanic skin activity results from the fact that the facilitatory region has largely been eliminated, and the inhibitory region becomes predominant. In our experiments, it was observed that in their respective distributions, facilitatory and inhibitory points were widely distributed and mutually interspersed. Although the method of stimulation did not distinguish which points were in cell masses and which ones were in conducting pathways, from the dispersion of facilitatory points in the pons and the medulla, as well as from the fact that after transection of the brain stem between the superior and inferior colliculi or posterior to the inferior colliculus, galvanic skin activity is restored following injection of nikethamide [11], it is apparent that in the reticular formation of the pons and the medulla, neuronal cell masses having a facilitatory effect can also exist, independently of descending fibers from the midbrain facilitatory center. The existence of the possibility of evoking a galvanic skin reflex by stimulation of ascending fibers cannot presently be entirely excluded, but in 3 animals, we removed the cerebrum anterior to the hypothalamus and then stimulated the midbrain reticular formation; the results thus obtained did not clearly differ from those obtained in other experiments in which the cerebrum was not removed.

As for the distribution of facilitatory and inhibitory centers in the brain-stem reticular formation, some workers [12] believe that the facilitatory and inhibitory region for the somatic system have similar effects on the respiratory and other vegetative functions. However, Bach [13], observing respiration and blood pressure simultaneously with the knee-jerk, noted that in the reticular formation the area of representation of each function was such that although all were well localized, stimulation of one point could frequently evoke all three responses. The direction and magnitude of the responses, however, were frequently different; points for which all three showed an increase constituted only 9% and the number of points showing a decrease of all three were still fewer. In our experiments, it was apparent that the distribution of points that facilitated or inhibited galvanic skin activity, respectively, were scattered and intermixed, without clearly delimited zones. The distributions are not at all the same as that of the somatic facilitatory and inhibitory regions delineated by Magoun et al. [14, 15]. At the same time, in the stimulation of more than a thousand individual points, those showing one type of response were the most frequent, constituting 56%, those showing two types amounted to 30%, and those showing all three types of responses amounted to only 14%. Among those points showing all three types of response, those resulting in parallel changes in galvanic skin activity, blood pressure, and respiration amounted to only 35%. For points showing two types of responses, the combination of only skin galvanic activity and respiratory responses or of blood pressure and respiratory responses, the responses themselves for the most part did not parallel one another; of the combination of only galvanic skin and blood pressure responses, in only 83% were the changes parallel ones. It is apparent that the facilitatory and inhibitory regions for different types of vegetative functions are not identical; each has its own mutually interspersed points, and only a portion are coincident. This type of juxtaposition or contiguity of anatomical sites is perhaps the basis of a close and fundamental relationship, and because of the extraordinarily complex, interlacing relationship between neurons, on the basis of the anatomical localization of noncontiguity or juxtaposition, it is impossible to reject the possibility that they have the capability of a close relationship.

CONCLUSIONS

1. With the aid of a stereotactic instrument and bipolar electrodes, the entire brain stem was systematically stimulated in these experiments; it was observed that points in the brain stem, the electrical stimulation of which resulted in galvanic skin activity, were of three types: (a) those which clearly altered galvanic skin activity and had a facilitatory effect on the gal-

vanic skin reflex, (b) points which had an inhibitory effect on the galvanic skin reflex, in that there was no clear galvanic skin response during stimulation, or the response developed slowly but after termination of stimulation a rebound response appeared, and (c) some points for which there was both a facilitatory and an inhibitory effect.

2. At the cephalic end of the midbrain reticular substance, points that facilitated galvanic skin activity predominated, but with gradual progression posteriorly, inhibitory points increased, and in the reticular formation of the medulla, the number of inhibitory and facilitatory points approximated one another. They were each distributed in a comparatively scattered manner, but were mutually interspersed; their distributions were not the same as the points of somatic facilitatory and inhibitory systems, respectively, which were delineated by Magoun and his collaborators.

3. With simultaneous recording of galvanic skin activity, blood pressure and respiratory responses, it was found that the points evoking a change in galvanic skin activity numbered the largest and were widely distributed. Points simultaneously evoking more than one of these types of responses, all having the same direction of change, were very few. The distribution of particular points capable of facilitation or inhibition, respectively, were dissimilar, although they were mutually interspersed, and only a small portion overlapped.

REFERENCES

1. Wang, G. H., The galvanic skin reflex: A review of old and recent works from a physiologic point of view, Am. J. Phys. Med., 36: 295-320 (1957); 37:35-37 (1958).
2. Wang, G. H., Brainstem reticular system and galvanic skin reflex, Fed. Proc., 14:158 (1955).
3. Wang, G. H., Stein, P., and Brown, V. W., Effects of transection of central neuraxis on galvanic skin reflex in anesthetized cats, J. Neurophysiol., 19:340-349 (1956).
4. Wang, G. H., Stein, P., and Brown, V. W., Brain reticular system and galvanic skin reflex in acute decerebrate cats, J. Neurophysiol., 19:350-355 (1956).
5. Wang, G. H., and Brown, V. W., Changes in galvanic skin reflex after acute spinal transection in normal and decerebrate cats, J. Neurophysiol., 19: 446-451 (1956).
6. Wang, G. H., and Brown, V. W., Suprasegmental inhibition of an autonomic reflex, J. Neurophysiol., 19: 564-572 (1956).
7. Wang, G. H., Strychninization of bulbar ventromedial reticular formation and galvanic skin reflex in thalamic cats, J. Neurophysiol., 21: 327-333 (1958).
8. Li Peng, Zheng Xiao-zhao, Jin Wen-quan, and Zhao Xiu-ju [Observations on the human galvanic skin potential and galvanic skin reflex] (in Chinese), Acta Physiol. Sinica, 25: 171 (1962).
9. Ni Guo-tan, Hui Zhao-lin, Zhang Ming-hua, and Zhu Hao-nian, [A simplified method for stereotactically implanting stimulation electrodes in the depth of the brain] (in Chinese), Progress in Physiol., 4:317 (1962).
10. Wang, G. H., and Brown, V. W., Terminal rebound of galvanic skin reflex in anesthetized cats, J. Neurophysiol., 20:340-346 (1957).
11. Li Peng, Cheng Jie-shi, and Sun Zhong-han, [Analysis of the mechanism of synchronization of galvanic skin activity and respiration] (in Chinese), Acta Physiol. Sinica, 28:378 (1965).
12. Hoff, H. E., Breckenridge, C. G., and Spencer, W. A., Suprasegmental integration of cardiac innervation, Am. J. Physiol., 171: 178-188 (1952).
13. Bach, L. M. N., Relationships between bulbar respiratory, vasomotor, and somatic facilitatory and inhibitory areas, Am. J. Physiol., 171: 417-435 (1952).
14. Magoun, H. W., and Rhines, R., An inhibitory mechanism in the bulbar reticular formation, J. Neurophysiol., 9: 165-171 (1946).
15. Rhines, R., and Magoun, H. W., Brain stem facilitation of cortical motor response, J. Neurophysiol., 9: 219-229 (1946).

LAMBDA WAVES OF HUMAN SUBJECTS OF DIFFERENT AGE LEVELS*

Tsai Hao-jan and Liu Shih-yih

THE PROBLEM

Lambda waves are a relatively recently discovered type of electrical activity of the human brain which so far has been little investigated systematically. Evans [3] in 1949 was the first to report so-called "sharp waves" in the occipital region, which in 1952 [4] he termed lambda waves. Lambda waves are recognized as being a type of electrical activity that is closely related to ordinary physiological and psychological phenomena. Under conditions of good illumination, during spontaneous visual scanning, lambda waves are clearly evident, but during steady fixation at a point, the waves are absent. They are also absent when the eyes are closed or under conditions of darkness [4, 5]. Cobb and Pampiglione [2], Roth and Green [12], Green [7], and Groethuysen and Bickford [8] have all observed these phenomena. In addition Gastaut and Bert [6] observed that the brain readily manifested lambda waves when a movie was being watched. Roth, Shaw, and Green [13] observed that lambda waves were readily apparent when K-complexes, which appear from the brain during sleep, were present.

At the present time, not much information is available concerning the mechanism of lambda waves. Some workers (Evans [4, 5]; Groethuysen and Bickford [8]; and Perez-Borja et al., [10]) consider that the region of the most prominent response to stimulation by light is frequently the same as that for lambda waves. Whether from the standpoint of physiology, psychology, electroencephalography, or clinical practice, we need new norms of the electrical activity of the brain; for this reason an investigation of lambda waves necessarily attracts our attention.

In our opinion, the majority of studies by previous workers have emphasized adults as subjects of investigation, whereas studies concerning the age characteristics of lambda waves, and also knowledge concerning lambda waves and different physiological and psychological conditions is so far comparatively meager. Consequently, in the present investigation the following two basic questions were examined: (1) the age characteristics of the appearance of lambda waves in the occcipital region, and (2) the relation between lambda waves and different physiological and psychological conditions.

*Acta Psychologica Sinica, No. 4, pp. 343-352 (1965).

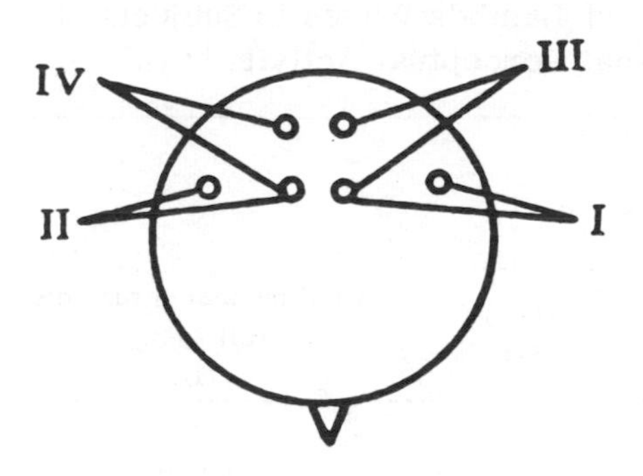

Fig. 1. Arrangements of leads for recording occipital lambda waves.

METHODS

A total of 71 subjects, 34 males and 37 females, were studied; all were ordinary healthy children, youths, adults, and elderly subjects, although the number of the latter was quite small. During an experiment, the subjects sat on a sofa in a shielded, soundproofed room, initially under conditions of darkness for several minutes at rest with the eyes closed, in order to record the spontaneous potentials of the background EEG. Afterwards, subjects were asked to open their eyes and to fixate directly ahead at a 1.5 × 1.0 cm neon bulb. Stimuli of single or repetitive light pulses (approximately 25 lux), obtained from a pulse generator, were used in order to observe the characteristics of the subjects' evoked potentials. In order to record lambda waves with the eyes open, good illumination is necessary, and hence, except when visual imagery was being studied, all of the experiments were performed with an electric light (60 W) in the soundproof room turned on. In the first part, the appearance of lambda waves was examined during the subjects' fixation on a point, scanning a white sheet of paper, and scanning a picture. Afterwards, in the majority of the subjects, the question of the appearance of lambda waves was examined during visual imagery consisting of imagined scanning of a picture, during mental arithmetic, or while making auditory discriminatory responses. Fixation at a point refers to subjects fixating at a neon bulb placed 20 cm directly ahead. Scanning a piece of white paper refers to subjects scanning a 50 × 80 cm white paper at a distance of 150 cm. Scanning a picture implies subjects scanning 50 × 80 cm colored pictures at a distance of 150 cm. Imagined scanning refers to subjects with eyes open in the dark room imagining scanning the aforementioned colored picture. The difficulty of the task of mental calculation selected was appropriate for the age and level of schooling of the subjects. For example, 3- to 7-year-old children mainly counted the sounds "Da," or mentally carried out addition (A + B); young subjects and adults mentally carried out single multiplication (AB × C) or double multiplication (AB × CD). The auditory discriminatory responses were made by subjects in response to auditory stimuli generated with the aid of a pulse generator and an electroacoustical transducer. Positive stimulus signals were 6/sec sounds and negative stimulus signals were 4-5/sec sounds. The appropriate response to the positive stimulus signals was for subjects to press a switch. The latter and the stimulus signals were automatically recorded on the EEG paper. The experimenter maintained contact with the subjects by means of an intercom system.

For the entire investigation, a 16-channel Ediswan-III was used for recording the EEG's. Since a high gain was required for recording lambda waves, a gain setting of 9 was used when recording the EEG's of adults, which corresponded to a voltage calibration of approximately 50 μV = 20 mm. For recording the EEG's of children, a gain of 8 was employed. The time-constant value was 0.3, and the high-frequency cutoff value was 15. In order to evaluate the localization characteristics of lambda waves, the study focused on recording of occipital potentials; bipolar recordings were used for each subject, transverse and longitudinal linkages on the left and right occiput, respectively, giving a total of 4 channels of potentials (see Fig. 1).

Table 1. Comparison of the Percentage of Occurrence of Lambda Waves in Subjects of Different Ages under Conditions of Visual Perceptual Activity*

Circumstances of appearance of lambda waves	Age in years (Number of subjects)					Total number of subjects (all ages) (71)
	3-7 (17)	9-12 (17)	18-30 (18)	31-50 (11)	61-80 (8)†	
Prominent	11 (64.7%)	10 (58.8%)	10 (55.6%)	3 (27.3%)	2	36 (50.7%)
Occasional	3 (17.6%)	4 (23.5%)	3 (16.6%)	1 (9.1%)	2	13 (18.3%)
Absent	3 (17.7%)	3 (17.7%)	5 (27.8%)	7 (63.6%)	4	22 (31.0%)

* Scanning of a well-illuminated picture.
† Percentages were not calculated for less than 10 subjects.

RESULTS

1. Analysis of Age Characteristics of the Appearance of Lambda Waves

The investigation showed clearly that among the 71 subjects of different ages, under conditions of visual perceptual activity, lambda waves were apparent in 49 subjects (69%) altogether. Of these the waves were prominent in 50.7% (36 subjects), and occasional in 18.3% (13 subjects). A prominent appearance of lambda waves under conditions of visual perceptual activity means that the waveform was clear, the periods during which they appeared were comparatively long, and the waves were frequently repetitive. Occasional lambda waves implies that although the waveforms were clear, the periods of their appearance were brief or the waves were not definitely repetitive. The appearance of lambda waves had at least the following two characteristics: (1) their rhythm varied, depending on the presence of other types of waves in the occipital background activity; (2) individual lambda waves were principally triphasic in form. In the transverse linkages, the waves were principally positive, i.e., the two medial leads in the occipital area were positive with respect to the lateral leads.

The results of the study make clear that there is an age dependence of lambda waves. Thus, in 3- to 7-year-old children and in 9- to 12-year-old youths, lambda waves were very prominent; they were also relatively prominent in adults of less than 30 years, but not very prominent in adults of over 30 years. From Table 1 it can be seen that among 3- to 7-year-old subjects there were altogether 14 subjects (82.3%) who showed lambda waves, in the 9- to 12-year-old group there were 14 subjects (82.3%), in the age group of 18 to 30 years, there were 13 subjects (72.2%), and in the 31- to 50-year-old group there were only 4 subjects (36.4%). Among the 8 subjects in the 61- to 80-year-old group there were only 4 who showed lambda waves, and in two of these, only occasionally. With respect to subjects having prominent lambda waves, the percentage of appearances of lambda waves in subjects under age 30 was more than twice that for subjects of ages over 30 years.

The age difference of lambda waves is distinguished not only by the dissimilarities in the percentage of appearance, but also by differences in wave form. There were, however, no age

Table 2. Comparison of Appearance of Occipital Lambda Waves in Different Age Groups

Age (number of subjects in parentheses)	Appearance of lambda waves in occipital linkages				Duration of lambda waves			Amplitude of lambda waves				Frequency of lambda waves
	Simult. present in four linkages in the two hemispheres	Present only in the two transverse linkages	Present only in the two A-P linkages	Present only in the two linkages of one hemisphere	< 0.1 sec	0.1-0.2 sec	0.2-0.3 sec	under 10 μV	10-20 μV	20-30 μV	30-50 μV	
3-7 (14)	7 (50.0%)	7 (50.0%)	0	0	0	6 (42.9%)	8 (57.1%)	0	1 (7.1%)	5 (35.8%)	8 (57.1%)	Children and youths: 2-3/sec as a rule
9-12 (14)	9 (64.3%)	5 (35.7%)	0	0	1 (7.1%)	13 (92.9%)	0	0	2 (14.3 %)	11 (78.6%)	1 (7.1%)	
18-30 (13)	5 (38.5%)	6 (46.2%)	0	2 (15.3%)	10 (76.9%)	3 (23.1%)	0	9 (69.2%)	4 (30.8%)	0	0	Adults: 3-5/sec as a rule
31-50 (4)*	3	1	0	0	4	0	0	3	1	0	0	
61-80 (4)*	3	0	1	0	3	1	0	0	4	0	0	
Total (49)	27 (55.1%)	19 (38.8%)	1 (2.0%)	2 (4.1%)	18 (36.7%)	23 (46.9%)	8 (16.4%)	12 (24.5%)	12 (24.5%)	16 (32.7%)	9 (18.3%)	

* Percentages not calculated for less than 10 subjects.

differences with respect to the localization of lambda waves. With respect to the frequency of lambda waves, their rhythm varied according to the prominence of other waves in the background of the occipital region, and hence the frequency of lambda waves was not the same as for the alpha, etc., and was rather unstable. However, the band of frequency variation was limited to 2 to 5/sec. Taking adults as an example, often at one moment the frequency was 3 to 4/sec, but at another moment it was 4 to 5/sec. Under still other conditions, the above-mentioned frequencies appeared to change, although for a few subjects the frequency was stable. It is interesting that there was a tendency of an increase of the frequency of lambda waves with increasing age. The frequency of lambda waves in children and youths was primarily approximately 2 to 3/sec, whereas for adults and elderly subjects, the frequency was primarily 3 to 5/sec.

A low amplitude is one of the characteristics of lambda waves as recorded from the scalp, and consequently it is very difficult to record lambda waves with the gains that are used for recording human alpha waves. When the EEG gain was increased to 50 μV/20 mm (principally for recording from adults) or to 50 μV/10 mm, then the amplitude of the lambda waves could be determined; for the most part they were under 30 μV, and only rarely exceeded 50 μV. The investigation showed that the amplitude of the lambda waves was comparatively high for children and youths, and comparatively low for adults. From Table 2 it can be seen that the amplitude of lambda waves of subjects of 3 to 7 years was principally 30 to 50 μV for 57.1% (8 subjects); for the 9 to 12 year group it was 20 to 30 μV as a rule for 78.6% (11 subjects). It is noteworthy that among the above-mentioned two groups there was no instance of lambda waves of an amplitude of less than 10 μV. The amplitude of lambda waves in adults was significantly lower; there was no instance in which it exceeded 20 μV. In subjects of 18 to 30 years, the amplitude of lambda waves was generally under 10 μV for 69.2% (9 subjects). Among the 4 subjects of ages 31 to 50 years, there were 3 who likewise had amplitudes of under 10 μV. Among the 4 elderly subjects (ages 61-80), the amplitude of lambda was 10 to 20 μV.

Since lambda waves are frequently not continuous in their appearance, and preceding and following lambda waves the EEG can be flat, or other waves can be present, then the duration of lambda waves cannot simply be considered as the reciprocal of the frequency. Generally, the age difference of the duration of lambda waves is clearer than the age difference for the frequency, i.e., the former is more regular in nature. Thus, lambda waves of children and adolescents are longer in duration than those of adults. From Table 2 it can be seen that there was not a single instance of lambda waves in children of duration of less than 0.1 sec, for 42.9% of subjects (6 individuals) the duration was 0.1 to 0.2 sec, and for 57.1% (8 individuals) the duration was 0.2 to 0.3 sec. Among the subjects 9 to 12 years old, there was only 1 example of lambda waves of duration of less than 0.1 sec, for the other 92.9% of subjects (13 individuals), the duration was 0.1 to 0.2 sec. For the large majority of adults, lambda waves were of less than 0.1 sec in duration. For example, among subjects of 18 to 30 years, 76.9% (10 persons) had lambda waves of less than 0.1 sec, the 4 persons in the 31 to 50 year group of subjects had lambda waves of less than 0.1 sec, and among the 4 persons in the 61 to 80 year group there were 3 with lambda waves of duration less than 0.1 sec. It is worth directing attention to the fact that for subjects of above 9 years of age, lambda waves did not exceed 0.2 sec in duration.

As was pointed out above, there was no age difference with respect to the localization of lambda waves, and fundamentally, lambda waves for subjects of different ages appeared approximately symmetrically in the two hemispheres. Moreover, the waves from the transverse linkages were more prominent than those from the antero-posterior linkages. From Table 2 it can be seen that among the 49 subjects of different ages who showed lambda waves, the latter appeared simultaneously in the four linkages on the two hemispheres in 55.1%, whereas they appeared in only the transverse linkages in 38.8%, and for only 1 person were they apparent in only the antero-posterior linkages of the two hemispheres. It should be pointed out that simultaneous appearance in the four linkages of the two hemispheres does not mean equally prominent

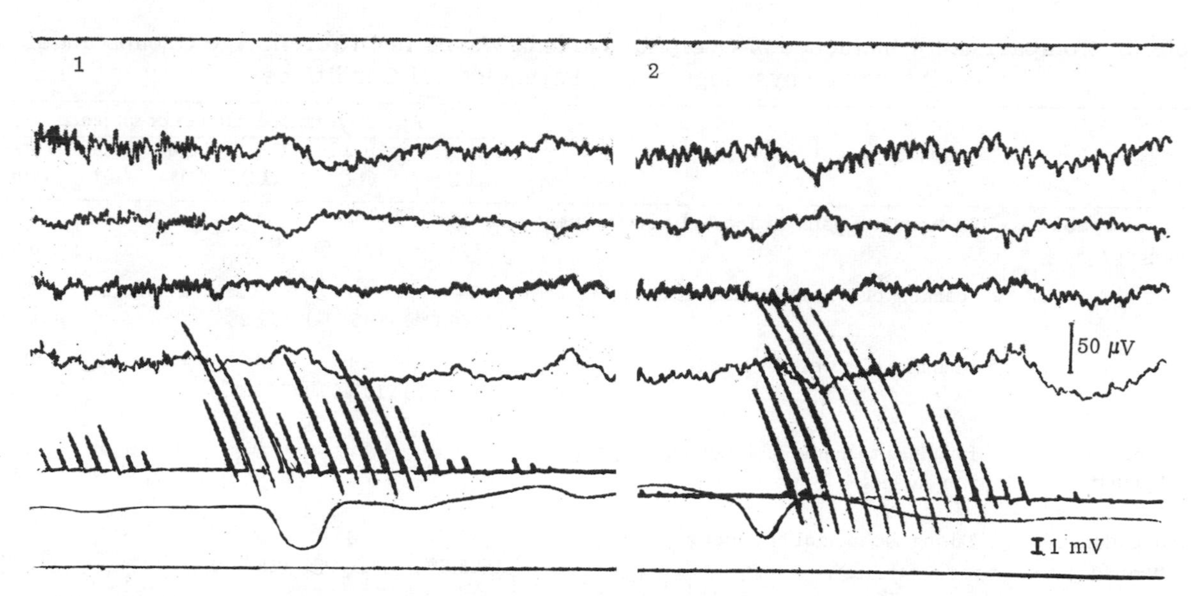

Fig. 2. Example of appearance of lambda waves in 3- to 7-year-old children while scanning a picture. ("Lambda" subject No. 51, 3-year-old female, recorded June 2, 1964). From top to bottom: Time markers (1/sec), left transverse occipital, right transverse occipital, left antero-posterior occipital, right antero-posterior occipital, frequency analysis of left transverse occipital activity, galvanic skin responses and stimulus channel. 1) Absence of lambda waves while fixating on a point. 2) Presence of lambda waves while scanning a picture.

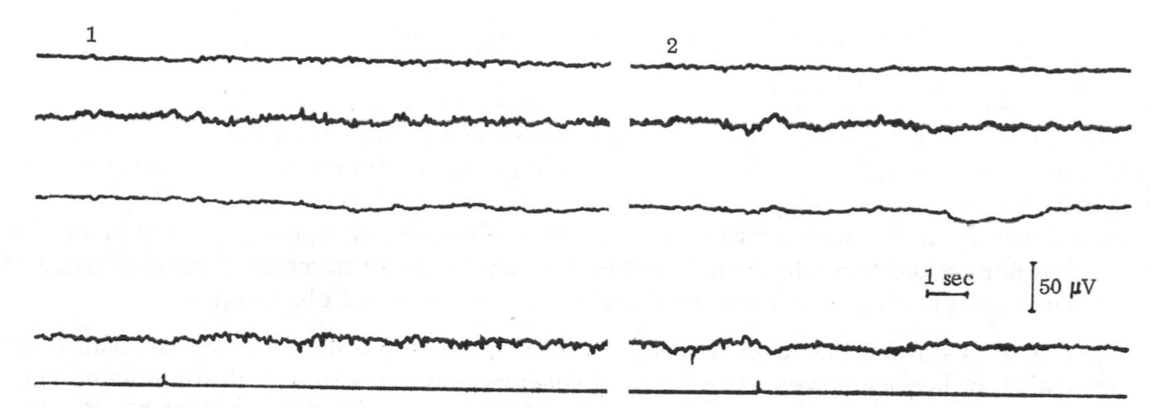

Fig. 3. Example of appearance of lambda waves in an adult while scanning a picture. ("Lambda" subject No. 38, 28-year-old male, recorded April 8, 1964.) From top to bottom: right transverse occipital, left antero-posterior occipital, right antero-posterior occipital, left transverse occipital, and stimuli, respectively. 1) Appearance of lambda waves while scanning a picture. 2) Absence of lambda waves during imagined scanning of the picture with eyes open in the dark.

appearance in the four linkages. Although lambda waves appeared almost always synchronously in the four linkages, it frequently happened that the waves in the transverse linkages in the two hemispheres were equally prominent, whereas in the antero-posterior linkages of the two hemispheres the waves were unequal in amplitude.

Table 3. Comparison of Incidence of Occipital Lambda Waves in Different Age Groups under Different Physiological and Psychological Conditions

	Conditions	Age in years and number of subjects					
		3-7 (14)	9-12 (14)	18-30 (13)	31-50 (4)*	61-80 (4)*	Total (49)
Visual perception (49 subjects)	Fixation at a point in a well-lit environment	5 (35.7%)	3 (21.4%)	0	0	0	8 (16.3%)
	Scanning of a well-lit sheet of paper	14 (100.0%)	12 (85.7%)	9 (69.2%)	1	1	37 (75.5%)
	Scanning of a well-lit picture	14 (100.0%)	14 (100.0%)	13 (100.0%)	4	4	49 (100.0%)
Visual imagery (42 subjects)†	Imagined scanning of a picture, eyes open in darkness	0	0	0	0	0	0
Mental arithmetic (42 subjects)†	During the mental arithmetic	2 (28.6%)	3 (21.4%)	2 (15.4%)	0	0	7 (16.7%)
Auditory discriminatory responses (42 subjects)†	During positive signalling stimuli	0	1 (7.1%)	2 (15.4%)	0	0	3 (7.1%)
	During negative signalling stimuli	0	2 (14.3%)	2 (15.4%)	0	0	4 (9.5%)

* See footnote to Table 2.

† 7 children, in the 3- to 7-year-old group could not carry out the mental calculation, auditory discriminative responses, or visual imagery.

2. Analysis of Lambda Waves under Different Physiological and Psychological Conditions

The question of the appearance of lambda waves under comparatively different physiological and psychological conditions (visual perception, visual imagery, mental arithmetic, and during differential responses to auditory stimuli) was also analyzed. The results showed that the appearance of lambda waves was closely related to visual perception. During mental calculation and during auditory differential responses, only a small number of subjects showed lambda waves, and during visual imagery, lambda waves were not observed.

Our results showed that with respect to visual activity, the most favorable conditions for the appearance of lambda waves were those of scanning a picture in a well-illuminated environment. With these conditions as a reference point, lambda waves appeared in 75.5% of subjects (37 individuals) while scanning the sheet of white paper in a well-illuminated environment. Analysis of the age distribution of the appearance of lambda waves under conditions of visual scanning of the white paper under bright illumination showed that lambda waves were very prominent in the 3- to 7-year-old group (100%), intermediate for the 9- to 12-year-old group (85.7%) and for the 18- to 30-year old group (69.2 %), and not at all prominent in the 31- to 50- and 61- to 80-year-old groups. While fixating at a point under conditions of bright illumination, there was no instance of the appearance of lambda waves in subjects of above 18 years. In the 3- to 7-year old and the 9- to 12-year-old groups, only 35.7%(5 individuals) and 21.4% (3 individuals), respectively, showed lambda waves. Figure 2 shows illustrative examples of fixating at a point and scanning a picture over a wide visual field, respectively. It is noteworthy that when subjects of different ages with eyes open imagined scanning the picture in darkness, lambda waves did not appear in a single instance. Figure 3 is an illustrative example of scanning a picture,

Table 4. Relation between Lambda Waves and the Type of Background Rhythm and the Phenomena of "On Responses" and "Driving Response," for Different Ages

Age in years (Number of subjects)	Conditions															
	Type of background rhythm								"On response"				"Driving response"			
	θ(α) or α(θ)		α		α(β) or β(α)		β		Present (53 subs.)		Absent (18 subs.)		Present (41 subs.)		Absent (30 subs.)	
	P*	A*	P*	A*	P*	A*	P*	A*	P*	A*	P*	A*	P*	A*	P*	A*
3-7 (17)	11 (64.6%)	2 (11.8%)	2 (11.8%)	1 (5.9%)	1 (5.9%)	0	0	0	13 (86.7%)	2 (13.3%)	1	1	12 (80.0%)	3 (20.0%)	2	0
9-12 (17)	0	2 (11.8%)	14 (82.3%)	1 (5.9%)	0	0	0	0	13 (86.7%)	2 (13.3%)	1	1	14 (87.5%)	2 (12.5%)	0	1
18-30 (18)	0	0	9 (50.0%)	3 (16.7%)	2 (11.1%)	2 (11.1%)	2 (11.1%)	0	12 (100.0%)	0	1	5	5	0	8 (61.5%)	5 (38.5%)
31-50 (11)	0	0	3 (27.3%)	4 (36.3%)	2 (13.2%)	1 (9.1%)	0	1 (9.1%)	3	3	1	4	2	0	2	7
61-80 (8)†	0	0	4	2	0	2	0	0	4	1	0	3	2	1	2	3
Total (71)	11 (15.5%)	4 (5.6%)	32 (45.2%)	11 (15.5%)	5 (7.0%)	5 (7.0%)	2 (2.8%)	1 (1.4%)	45 (84.9%)	8 (15.1%)	4 (22.2%)	14 (77.8%)	35 (85.3%)	6 (14.7%)	14 (46.7%)	16 (53.3%)

*P — lambda waves present; A — lambda waves absent.

† See footnote to Table 2.

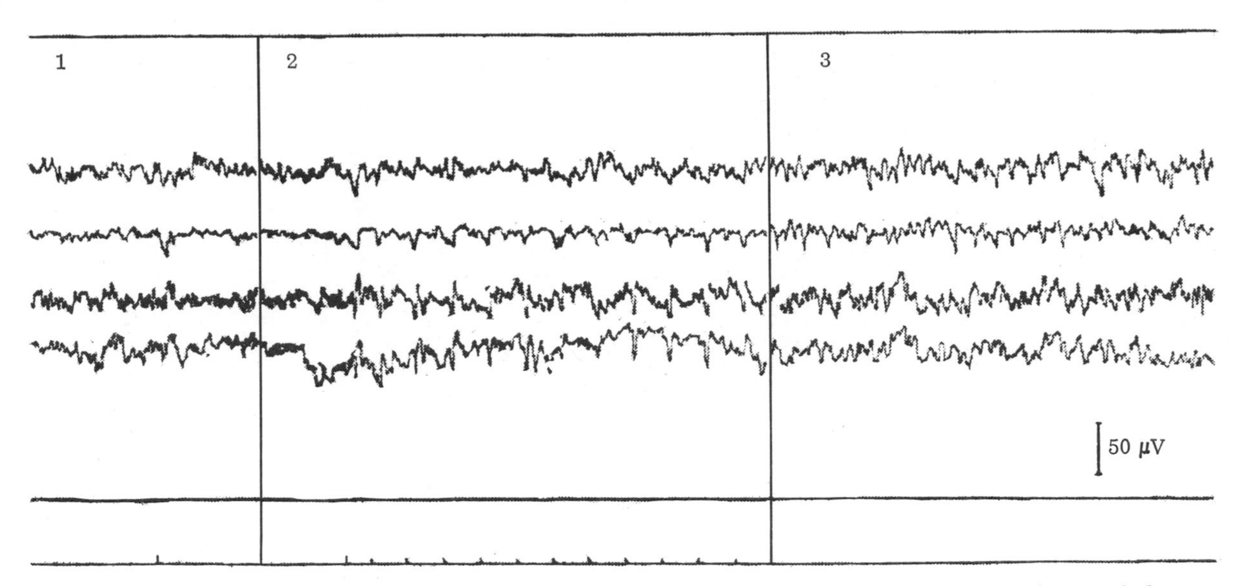

Fig. 4. Example of the basic similarity of polarity of lambda waves and those of the phenomena of "on responses" and the "driving response." ("Lambda" subject No. 24, 11-year-old female, recorded March 27, 1964.) From top to bottom: time markers (1/sec), left transverse occipital, right transverse occipital, left antero-posterior occipital, right antero-posterior occipital, blank channel, and stimuli, respectively. 1) "On-response" evoked by a single flash of light. 2) "Driving response" evoked by repetitive flashes of light. 3) Appearance of lambda waves during scanning of a picture.

and imagined scanning of the picture. From investigation of mental calculation and auditory discriminative responses, it was apparent that lambda waves also mainly appeared in the group of subjects of less than 30 years of age. From Table 3 it can be seen that during mental arithmetic only 16.7% of subjects (7 individuals) showed lambda waves, and while discriminatory responses were being made to positive and to negative signalling auditory stimuli, only 7.1% (3 individuals) and 9.5% (4 individuals) showed lambda waves.

We now further analyze for different ages the question of the relation of lambda waves to the type of background rhythm and to the phenomena of the "on response" and the "driving response" ("assimilation of rhythm"), since these are appropriate questions for delineating the nature of lambda waves. From the relation between lambda waves and the type of background brain rhythms, the results of the investigation showed the following: all subjects having an occipital background rhythm of predominantly theta with some alpha [$\theta(\alpha)$], or alpha with some theta [$\alpha(\theta)$], or alpha alone [α], readily showed lambda waves, but in subjects having an occipital background rhythm of predominantly alpha with some beta [$\alpha(\beta)$], or predominantly beta with some alpha [$\beta(\alpha)$], or primarily beta [β], lambda waves were relatively infrequent. From Table 4, it is apparent that among 58 subjects having $\theta(\alpha)$, $\alpha(\theta)$, or α types of background rhythms, the number of subjects (43) showing lambda waves was almost three times the number (15) of those not showing lambda waves. However, among 13 subjects having $\alpha(\beta)$, $\beta(\alpha)$, or β types of background rhythms, the number of those showing lambda waves (7) was about equal to the number of those not showing them (6).

The investigation made clear that lambda waves are closely related to single evoked potentials (the phenomenon of the "on-response") and to repetitive evoked potentials (the phenomenon of the "driving response"). For subjects showing occipital "on responses" and "driving responses," lambda waves could readily be recorded under conditions of visual perception.

Under almost all conditions, the phase (polarity) of lambda waves was the same as that for the "on response" and particularly for the "driving response."

The relation between lambda waves and "on responses" will be analyzed first. From Table 4 it can be seen that among 53 subjects showing occipital "on-responses" to varying degrees, lambda waves were recorded from 84.9% (45 subjects), but among 18 subjects who did not show the phenomenon of "on responses," in only 22.2% (4 subjects) were lambda waves recorded. Under most circumstances, the polarity of lambda waves was the same as that of the "on responses." With respect to the study of EEG activity, we know that for the same subject, the polarity of the "on response" is not constant, but under the conditions of the present investigation, at least for the two transverse linkages, the polarity was constant.

The relation between lambda waves and the "driving response" will next be analyzed. From Table 4 it can be seen that among 41 subjects showing an occipital "driving response" to varying degrees, lambda waves could be recorded from 85.3% (35 subjects), but among 30 subjects who did not show the "driving response," lambda waves could be recorded from only 46.7% (14 subjects). It is worth noting that for almost all subjects the polarity of lambda waves was the same as that of the "driving response" (see Fig. 4).

DISCUSSION

The general nature of normal lambda waves is a question of interest to investigators of the electrical activity of the brain. The percentages of occurrence of normal lambda waves that have been reported by previous workers are quite diverse. For example, Evans [5] reported that only 2% of subjects showed lambda waves, Roth and Green [12] reported 60%, and Rémond and Lesèvre [11] reported that in 61-99% of subjects, lambda waves were apparent. Mundy-Castle et al. [9] reported that among 50 normal elderly subjects of an average age of 75.1 years, only 2 subjects showed lambda waves. The present investigation makes clear that the appearance of normal lambda waves is related to the age of the subjects. It was apparent that their appearance was evidently inversely related to the presence of beta waves: lambda waves were very prominent in children and youths, and they gradually became less prominent with increasing age; in children and youths, beta waves were not at all prominent, but they also gradually become more prominent with increasing age. The age dependence of the change of wave form characteristics was very similar to those for alpha waves; and if the age group above 60 years with special characteristics of the appearance of alpha waves is not included, then the latter also shows an increase with increasing age. The amplitude and duration of lambda waves, however, show a tendency of diminution with increasing age.

The investigation of the relation of lambda waves to different physiological and psychological conditions is an interesting problem. Our study, on the one hand, confirmed that there is indeed a close relation between the appearance of lambda waves and visual perception, and on the other hand, it was also shown that under other physiological and psychological conditions, lambda waves could also appear, although they were less prominent. Roth and Green [12] reported that lambda waves could be seen during mental calculation of a difficult problem. Roth et al. [12, 13] reported that, during light sleep or under light barbiturate anesthesia, lambda waves were also present or even increased. Recently, in an investigation concerning the human EEG during sleep [1], we also observed a certain number of lambda waves in a few subjects during stages B and C of sleep. The fact that lambda waves could appear during sleep leads us to conclude that a well-lit environment and open eyes are not essential conditions for the appearance of lambda waves, although under waking conditions the previously mentioned relationships are correct. However, we found that during scanning of a picture, lambda waves were more prominent than during scanning of a large sheet of white paper. Finally, during visual imagery of the picture with the eyes open in darkness, not a single subject showed lambda waves.

In relation to the mechanism of lambda waves, Rémond and Lesèvre [11] point out that lambda waves are comparatively infrequently recorded from subjects having a regular alpha rhythm, but for other subjects having more complex and not well-synchronized rhythms, lambda waves often appear. Our results make clear that among subjects of different ages, for background rhythms of theta with some alpha, alpha with some theta, or alpha types, lambda waves are encountered almost three times more frequently than not. However, for subjects with background rhythms of alpha with some beta, beta with some alpha, or beta types, the number of subjects showing lambda waves was almost equal to the number of those who did not. It is noteworthy that lambda waves and single evoked occipital electrical potentials (the phenomenon of the "on response") and repetitive evoked electrical potentials (the phenomenon of the "driving response") are related questions. Perez-Borja et al. [10], by means of a gross recording electrode at the time of surgery on a patient, observed that the localization of lambda waves was the same as that for "on responses" and "off responses." The present investigation further makes clear that, in the main, when "on responses" and/or the "driving response" were present in the occipital region, lambda waves were readily apparent under conditions of visual perception, and that the polarity of lambda waves was the same as that for "on responses" and for the "driving response." The above-mentioned results indicate that the mechanism of lambda waves is obviously closely related to that of evoked potentials, and consequently lambda waves can only be a manifestation of the electrical activity of the cerebral cortex.

CONCLUSIONS

1. The appearance of lambda waves was most marked in children and youths, less marked in 18- to 30-year-old adults, and least marked in ages over 31 years.

2. With respect to wave form, the frequency of lambda waves was higher in adults than in children and in youths, but the amplitude and duration in adults were less than in children and youths.

3. The appearance of lambda waves was closely related to visual perception (very prominent during scanning of a picture, and less so during scanning of a sheet of white paper), but during visual imagery (imagined scanning of a picture with eyes open in darkness), lambda waves did not appear.

4. If the phenomena of occipital "on responses" and/or "driving responses" were present, lambda waves were very easily recorded during visual perception.

REFERENCES

1. Wu Qin-e, Wu Zhen-yun, and Liu Shih-yih, [The EEG during sleep in human subjects of different age levels] (in Chinese) (1965) (unpublished).
2. Cobb, W. A., and Pampiglione, G., Occipital sharp waves responsive to visual stimuli, Electroenceph. clin. Neurophysiol., 4:110-111 (1952).
3. Evans, C. C., Comments on occipital sharp waves responsive to visual stimuli, Electroenceph. clin. Neurophysiol., 4:111 (1952).
4. Evans, C. C., Some further observations on occipital sharp waves (λ waves), Electroenceph. clin. Neurophysiol., 4:371 (1952).
5. Evans, C. C., Spontaneous excitation of the visual cortex and association areas — lambda waves, Electroenceph. clin. Neurophysiol., 5:69-74 (1953).
6. Gastaut, H. J., and Bert, J., EEG changes during cinematographic representation, Electroenceph. clin. Neurophysiol., 6:433-444 (1954).
7. Green, J., Some observations on lambda waves and peripheral stimulation, Electroenceph. clin. Neurophysiol., 9:691-704 (1957).

8. Groethuysen, U. C., and Bickford, R. G., Study of the lambda wave response of human beings, Electroenceph. clin. Neurophysiol., 8:344 (1956).
9. Mundy-Castle, A. C., Hurst, L. A., Beerstecher, D. M., and Prinsloo, T., The electroencephalogram in the senile psychoses, Electroenceph. clin. Neurophysiol., 6:245-252 (1954).
10. Perez-Borja, C., Chatrian, G. E., Tyce, F. A., and Rivers, M. H., Electrographic patterns of the occipital lobe in man: A topographic study based on use of implanted electrodes, Electroenceph. clin. Neurophysiol., 14:171-182 (1962).
11. Rémond, A., and Lesèvre, N., The conditions of appearance and the statistical importance of the lambda waves in normal subjects, Electroenceph. clin. Neurophysiol., 8:172 (1956).
12. Roth, M., and Green, J., The lambda waves as a normal physiological phenomenon in the human EEG, Nature, 172:864-866 (1953).
13. Roth, M., Shaw, J., and Green, J., The form, voltage distribution, and physiological significance of the K-complex, Electroenceph. clin. Neurophysiol., 8:385-402 (1956).

ELECTROENCEPHALOGRAPHIC AND GALVANIC-SKIN INVESTIGATION OF THE ORIENTING REFLEX IN MAN*

Liu Shih-yih and Wu Qin-e†

The term "orienting reflex" was adopted as early as 1910 by Pavlov, and in the paper "Internal inhibition of conditioned reflexes and sleep—one and the same process," published in 1923 [9], he defined the term in detail. The Pavlov school carried out important studies of the role of the orienting response in the formation of conditioned reflexes, for example, establishing that the orienting reflex is of major importance in the establishment of temporary connections to indifferent stimuli (Podkopaev [10]; Narbutovich and Podkopaev [7]). In these classical investigations, the orienting reflex was not itself studied independently or specifically, and therefore its structure and neural dynamic characteristics and other aspects have not so far been examined systematically. In the classical studies, the motor components (eye movements, head movements, etc.), respiratory and other vegetative components of the orienting reflex were principally studied; other components of the orienting reflex, however, were studied very little.

In the last ten years, the orienting reflex has been investigated extensively in two major respects: on the one hand, the orienting reflex has been taken as an independent problem for study of higher nervous activity, and, on the other hand, recent electrophysiological techniques have increasingly been used for studying directly the central components as well as the efferent vegetative components of the orienting reflex (Novikova and Sokolov [8]; Sokolov [11, 12]; Voronin and Sokolov [27]; Sokolov and Paramonova [14]; Jouvet [20]; Grastyān [17]; and others). As Anokhin [2] pointed out at a conference on special problems of the orienting reflex, the fundamental theoretical problem of the orienting reflex is very much debated at the present time. Most workers agree that the orienting reflex is a type of unconditioned reflex. Thus, Asratian [3, 4] considers that a conditioned reflex results from the combination of two unconditional stimuli, of which one is the orienting reflex. However, Grastyān believes that the orienting reflex cannot be an unconditioned response, but is instead an expression of a conditioned reflex. This controversy is very closely related to the problem of the neural basis of the orienting reflex, i.e., whether the orienting reflex is principally a function of the cortex, or of subcortical structures. Recently, at the special conference in the Soviet Union on "Neural Mechanisms of Formation of Temporary Connections" (the 1960 Gagra Symposium), Voronin and Sokolov [5] asserted that the orienting reflex is principally a function of the cortex, but this view was contested by a number of other participants. We believe that further experimental studies related to this question would be helpful in elucidating the nature of the orienting response.

The orienting reflex is a generalized, diffuse, nonspecific, adaptative response. It can be evoked by any kind of external agent, and very frequently includes the activity of many reflex

---

*Acta Psychologica Sinica, No. 1, pp. 1-10 (1963).

†Cheng Su-zhen participated in this work.

arcs simultaneously. The orienting reflex can be divided into afferent, central, and efferent (motor and vegetative) components. There are many possibilities for the afferent component. Light, sound, and cutaneous stimuli are frequently used alone, generally with a duration exceeding one second of continuous or intermittent stimulation. Recently, some workers (Sokolov and Paramonova [15]; Voronin and Sokolov [27]) have suggested that there is an ideal "neural model" for the investigation of the afferent component of the orienting reflex; we agree with this view, but heretofore no one has used time intervals of less than half a second for patterns of pulsed single stimuli for investigating such a "neural model."

As an indicator of the central component, changes of the electrical activity of the brain (including the phenomenon of "desynchronization" with alpha inhibition) are a very sensitive indicator of the orienting reflex, since it extinguishes very slowly and rapidly reappears (Motokawa [23]; Segundo et al. [25]). There are numerous indicators for the efferent component, such as skin potential, eye movement, respiration, the EKG, and the EMG. According to previous workers (Voronin and Sokolov [5]; and others), as well as our own experiments, eye movements, respiration, the EKG, and the EMG are not all equally sensitive components of the orienting reflex.

In our studies, we have principally used patterns of pulsed single stimuli as afferent components for the orienting reflex, and we have used the EEG and skin potential as the most sensitive indicators for analysis of the central and efferent components, respectively, of the orienting reflex. The aim was to investigate the following three theoretical problems: (1) analysis of the central (EEG) component of the orienting reflex; (2) analysis of the efferent (skin potential) component of the orienting reflex; and (3) the problem of the unconditioned versus the conditioned orienting reflex.

METHODS

There were three parts of the study: in the first, repetitive stimulus patterns were initially used, and then patterns of single stimuli, to examine the problem of the central (EEG) component. In the second part, patterns of repetitive stimuli were also used initially, then patterns of single stimuli were employed, to study the efferent (skin potential) component of the orienting reflex. In the third part, single light and cutaneous stimuli were used initially to study the question of the unconditioned orienting reflex and then subsequently, to examine the problem of the conditioned orienting reflex.

The subjects were normal children, young people, and adults; the majority were elementary, intermediate, and high school students, together with a small number of personnel from organizations. In the first and second parts of the experiments, there were altogether 294 subjects, among whom in 150 the EEG alone was recorded, in 144 the EEG and skin potentials were recorded simultaneously, EKG's and EMG's also being recorded simultaneously in a part of these subjects. In the third part of the experiments, there were 26 subjects altogether, children and adults. The EEG, skin potentials, and EMG were all recorded on a 16-channel Ediswan Model III ink-writing electroencephalograph. For the EEG's bipolar recordings of occipital potentials were principally used, sometimes the temporal, parietal, frontal, and fronto-parietal potentials were also recorded, one ear being used as the indifferent and as the ground electrode. The skin potentials were recorded by Tarkhanov's method (1889); the potential difference was recorded between the palmar and dorsal surfaces of the left hand, using $2\frac{1}{2}$-cm-diameter laminated circular nonpolarizable silver electrodes applied with special electrode paste. In the second part of the experiments, the electrodes were applied at the same time to the palmar and dorsal surfaces of the left hand, right hand, and right leg in 30 subjects, the "dorsal — palmar" skin potential responses of which were connected to the preamplifier stage of three separate channels of the electroencephalograph, respectively, for recording. For the EKG recordings,

Table 1. Comparison of the Appearance of the Central (EEG) Component of the Orienting Reflex for a Repetitive Sensory Input (Light Stimuli) in Normal Subjects of Different Ages

Subjects		Alpha waves prominent during stimulation*	Alpha waves inhibited during stimulation*
Type	Number		
Children (4-6 years)	41	75.6% (31)	24.4% (10)
Youths (12-15 years)	50	4% (2)	96% (48)
Adults (20-40 years)	49	0	100% (49)

* Percentage and number of subjects (in parentheses).

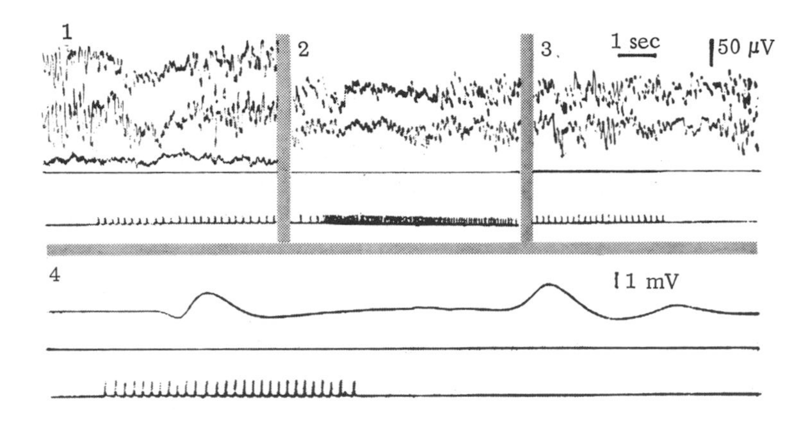

Fig. 1. Complex process of extinction of the orienting reflex evoked by continuous or interrupted repetitive sensory input. 1) At the moment of onset of the stimuli, "desynchronization" occurs, followed by very rapid adaptation (Subject No. C692, Jin X., 6-year-old female, recorded Sept. 9, 1960.) From top to bottom: right occipital, left occipital, frontal EEG's, single flashes. 2,3) Change of stimulus frequency (2) and cessation of stimulation (3) evoke "desynchronization" a second time. (Subject No. C846, Cha X., 4-year-old male, recorded Sept. 30, 1960.) From top to bottom: EEG's of right occiput, left occiput, flashes. 4) Appearance of a galvanic skin response (GSR) at onset of stimulation, which then adapts; following cessation of stimulation, a GSR appears a second time. (Subject No. C62, Shen X., 13-year-old male, recorded Sept. 22, 1962.) From top to bottom: Electrodermogram (skin potential) of right hand, flashes.

nonpolarizable silver electrodes were applied with the aid of electrode paste to the left and right wrists, respectively. For the EMG's, silver electrodes were used to record from the flexors of the right index finger. The equipment for the light and sound stimuli was constituted of a "pulse generator" and "electrolight" and "electroacoustic" transducers of our own manufacture. The pulse generator could produce single or repetitive short pulses of variable interval (frequency), duration (width), and amplitude, the single or repetitive brief pulses being used to trigger the "electrolight" and "electroacoustic" transducers. In the present investigation, comparatively weak stimulus strengths were used: 25 lux for the light and 25 db for the sound. For the cutaneous stimuli, the mechanism for cutaneous stimulation of a standard Pavlov experimental chamber was used, the stimuli being applied to the inner aspect of the left leg of the subjects.

When patterns of single stimuli were used, the trigger pulses for light and sound stimuli were both limited to a maximum duration of $^1/_{15}$ sec, although in practice the duration of the stimuli was slightly longer than that of the trigger pulses, although less than $^1/_2$ sec in any event. The duration of the cutaneous stimuli was 1 sec or less. During the experiments, the subjects sat on a sofa in a sound-insulated, electrically shielded darkened chamber, contact being maintained with the experimenter by means of a two-way intercom.

RESULTS

1. Analysis of the Central (EEG) Component of the Orienting Reflex

Since the orienting reflex is the reflection of a mechanism of "exploratory" activity (Anokhin [2]), it should doubtless be studied with sensory afferents that are basically related to the environment. We believe that the simplest, fundamental, sensory afferent is the basis of all complex sensory afferents, and hence investigation of this "model" of simplest form is of important theoretical significance. When we examined the characteristics of the central (EEG) component of the orienting reflex evoked by a repetitive sensory afferent, we found that an important deficiency of the latter was that the stimulus itself resulted in adaptation and extinction; this deficiency was especially apparent in children less than seven years old. From Table 1 it can be seen that 75.6% of 4- to 6-year-old children showed this phenomenon of adaptation during repetitive light stimuli, the alpha waves reappearing almost immediately. For this reason, the same environmental afferent which initially produced the central (EEG) component of the orienting reflex also resulted in its extinction. From Fig. 1, it can be seen that the onset of a repetitive light stimulus immediately results in "desynchronization," but subsequently there is rapid adaptation [Fig. 1(1)]; and immediately the stimulus frequency is changed, or the stimulation is terminated, "desynchronization" is again momentarily evoked a second time [Fig. 1(2) and 1(3), respectively]. These results demonstrate that the onset, change, or termination of a sensory input are all novel agents for the child, but if an input is sustained, a complex process of extinction of the orienting reflex occurs.

Traditionally, the central (EEG) component of the orienting reflex has referred only to the phenomenon of "desynchronization," but in actuality there are numerous changes. It is evident that single sensory stimuli can have at least the following two types of components in the human EEG: (1) the phenomenon of "desynchronization" and (2) an "on response" [Fig. 2(1) and 2(2), respectively]. Both are evoked by novel external stimuli, and the responses are always generalized and diffuse initially, although the former is most prominent in the occipital region, less so in the temporal region, whereas the latter is most easily evoked by sound stimuli in the incipient stages of sleep. The appearance of the above-mentioned two components is clearly related to the pattern of the spontaneous EEG activity; for example, if the alpha-index is high, there is prominent "desynchronization," and if the amplitude level is low, an "on response" is readily apparent, etc. However, there need not necessarily be a relation to the spontaneous activity; for example, an "on response" can be apparent if the alpha-index is high, and an "on response" is not necessarily apparent if there is a low-amplitude level, etc. The important point is that the above-mentioned two components are closely related to the significance for the subject of the input at the particular time. From this standpoint, it can be stated that the "on response" is frequently a relatively strong reflection of the orienting reflex, or, in other words, it is closely related to the arousal response, because the latter is very easily produced in subjects under conditions in which they are completely inattentive. Thus, the more unexpected the input, the greater is the likelihood of an "on response" and the greater its amplitude; contrariwise, the less the "unexpectedness" of the input, the less the likelihood of an "on response," or the lower its amplitude.

Table 2. Comparison of the Appearance of the Central (EEG) Component of the Orienting Reflex for a Single Sensory Input (Light Flash) in Normal Subjects of Different Ages

Subjects		Presence of "desynchronization"*		Presence of "on response"*	Simultaneous appearance of "desynchronization" and "on response"*	No response*
Type	Number	< 2 sec	> 3 sec			
Children (4-6 years)	50	56%(28)	2%(1)	0	0	42%(21)
Youths (12-15 years)	50	18%(9)	56%(28)	6%(3)	18%(9)	2%(1)
Adults (20-40 years)	50	4%(2)	70%(35)	10%(5)	14%(7)	2%(1)

* Percentage and number of subjects (in parentheses).

Table 3. Summary of the Appearance of the Central Component (Phenomenon of "Desynchronization)" of the Orienting Reflex for Single Sensory Inputs of Different Types in Normal Adults (20-40 years) and in Children (4-6 years)

Sensory modality	Adults (20-40 years)		Children (4-6 years)	
	First response	Rapidity of extinction	First response	Rapidity of extinction
Light	+++*	---*	+	–
Sound	+	–	+	–
Cutaneous	++	–	+	–

* The number of symbols shown is proportional to the degree of prominence of the first response, and of the difficulty of extinction of the response, respectively.

Table 4. Comparison of the Appearance of the Efferent (Skin Potential) Component of the Orienting Reflex for a Repetitive Sensory Input (Light Stimuli) in Normal Subjects of Different Ages

Subjects		Appearance of a phasic GSR upon initiation of stimulation*	Appearance of a phasic GSR upon cessation of stimulation*
Type	Number		
Children (4-6 years)	8	100% (8)	75% (6)
Youths (10-17 years)	45	100% (45)	57.7% (26)
Adults (20-40 years)	8	100% (8)	37.5% (3)

* Percentage and number of subjects (in parentheses).

When we analyzed the experimental results of 150 subjects of different ages, we found that an input of single sensory stimuli could result in the manifestation of "desynchronization" and/or an "on response" in 98% of young people and adult subjects, but in only 58% of children. From Table 2 it is apparent that in the large majority of children, the "desynchronization" did not persist for more than 2 sec, and as the age increased, the proportion of subjects for whom the duration of the "desynchronization" exceeded 3 sec clearly increased. This result demonstrates that the central (EEG) component of the orienting reflex in children is relatively weak, but it gradually becomes stronger with increasing age.

We found that the characteristics of the "desynchronization" of the central (EEG) component of the orienting reflex evoked by single sensory inputs of different types (weak light, sound, or cutaneous stimuli) were different with respect to the appearance of the response itself and with respect to its extinction. Thus, generally speaking, in adults the sensory input most strongly evoking the phenomenon of "desynchronization" was light stimuli, then cutaneous, and then sound stimuli. The extinction of "desynchronization" was the most prominent for sound stimuli, intermediate for cutaneous stimuli, and the smallest effect occurred for light stimuli. The selectivity for different types of sensory inputs for children of less than seven years differed from that for adults, as was apparent in their weaker "desynchronization" response and its especially ready extinction for the several different sensory inputs (see Table 3).

2. Analysis of the Efferent Component (Skin Potential) of the Orienting Reflex

When we carried out an analysis of the characteristics of the efferent component (skin potential) of the orienting reflex evoked by repetitive sensory stimuli in 61 subjects, we found that a major drawback of such stimuli was that the stimuli themselves resulted in an adaptation and extinction, the deficiency being apparent for subjects irrespective of age. Table 4 indicates that 37.5 to 75% of subjects showed the phenomenon of adaptation while being stimulated with repetitive light stimuli; when the stimulation was terminated, a phasic skin potential response (GSR) was evoked for the second time, very rapidly returning to the level existing before the phasic change. This result shows that the same external stimulus that had already evoked the efferent (skin potential) component of the orienting reflex also resulted in its extinction. In Fig. 1(4), it can be seen that the onset of the light stimulation evokes a GSR, after which there is very rapid adaptation, and cessation of the stimulus evokes a GSR for a second time. This result shows that the initiation of the input as well as its cessation are both novel for the subject, but that during the continuation of the input a complex process of extinction of the orienting reflex occurs.

We now consider the problem of the characteristics of the efferent (skin potential) component of the orienting reflex evoked by single sensory inputs. Traditionally, for investigating the efferent (skin potential) component of the orienting reflex, the GSR of only one hand was recorded. When we carried out recordings on 90 subjects, we found that all subjects did not show the GSR equally well; whether it appeared or not was closely related to the dorso-palmar resistance of the different subjects. For the Ediswan Model III apparatus, a resistance of 10,000 to 30,000 ohms was most suitable; the GSR was much less likely to be apparent if the resistance were less than 5000 ohms or greater than 50,000 ohms. The tonic GSR (the spontaneous slow skin potential change) is weak in adults but it is very prominent in children. From Table 5 it is apparent that a tonic skin response was apparent in only 18.9% of adults, whereas for children the percentage was much higher, as much as 95.5%. The phasic GSR was also weak in adults, whereas it was prominent in children. Under most conditions, in adults and young people the first phasic GSR evoked by single cutaneous stimuli was biphasic or triphasic, but those evoked by light or sound stimuli were monophasic. For children, however, the conditions were different, since the first sensory input of the different types all evoked a biphasic or triphasic skin-potential response relatively readily.

In order to investigate further the character of the efferent (skin potential) component of the orienting reflex, we recorded simultaneously the GSR for the right hand, left hand, and right leg from 15 adults and 15 children, which further affirmed that the efferent components of the orienting reflex are very prominent in children. We observed that except for certain conditions, a given sensory input evoked the same phasic GSR in the left and right hands, with the same latency, phase, and amplitude. However, the GSR in the upper and lower extremities were somewhat different. The latency of the GSR evoked by a given sensory input in the upper extremity was generally $\frac{1}{2}$ to 1 sec less than, as well as clearer than, that for the lower extremity. Whereas the GSR in the upper extremity is biphasic, the response in the lower extremity is mostly monophasic. The difference in the phases of the GSR's of the hand and foot were often

Table 5. Comparison of the Appearance of the Efferent Component (Skin Potential) of the Orienting Reflex for Single Sensory (Sound, Light, and Cutaneous) Stimuli among Normal Subjects of Different Ages

Subjects				Phasic GSR for first stimulus			Conditions of adaptation†			
				Analysis of waveform						
Type	Number	Single stimuli	Number of subjects responding	Monophasic	Biphasic	Triphasic	Easily extinguished	Extinction relatively difficult	Extinction difficult	Subjects showing tonic GSR
Children (4-6 years)	22	sound	12	50% (6)*	16.7% (2)	33.3% (4)	33.3% (4)	58.3% (7)	8.4% (1)	95.5% (21)
		light	16	6.3% (1)	12.5% (2)	81.2% (13)	0	18.8% (3)	81.2% (13)	
		cutaneous	17	23.5% (4)	17.6% (3)	58.9% (10)	0	11.8% (2)	88.2% (15)	
Youths (10-15 years)	31	light	24	62.5% (15)	25% (6)	12.5% (3)	58.4% (14)	41.6% (10)	0	58.1% (18)
		cutaneous	24	29.2% (7)	29.2% (7)	41.6% (10)	8.4% (2)	50% (12)	41.6% (10)	
Adults (20-40 years)	37	sound	8	62.5% (5)	25% (2)	12.5% (1)	50% (4)	37.5% (3)	12.5% (1)	18.9% (7)
		light	15	53.3% (8)	20% (3)	26.7% (4)	80% (12)	20% (3)	0	
		cutaneous	36	47.2% (17)	11.1% (4)	41.7% (15)	50% (18)	38.9% (14)	11.1% (4)	

* In each instance the lower number (in parentheses) is the number of subjects.

† "Easily extinguished" indicates that extinction had occurred after three trials; "extinction relatively difficult" indicates that the waveform had simplified with the third trial but the response was still present; "extinction difficult" indicates that with the third trial the response was still clear and without change in waveform.

Table 6. Simplified Summary of the Efferent (Skin Potential) Component Manifestation of the Orienting Response for Different Types of Single Sensory Inputs for Normal Adults (20-40 years) and Children (4-6 years)

Single sensory input	Adults (20-40 years)		Children (4-6 years)	
	Initial response	Rapidity of extinction	Initial response	Rapidity of extinction
Light	++*	–	+++	– – –
Sound	+	–	++	– –
Cutaneous	+++	– – –	+++	– – –

* See footnote for Table 3.

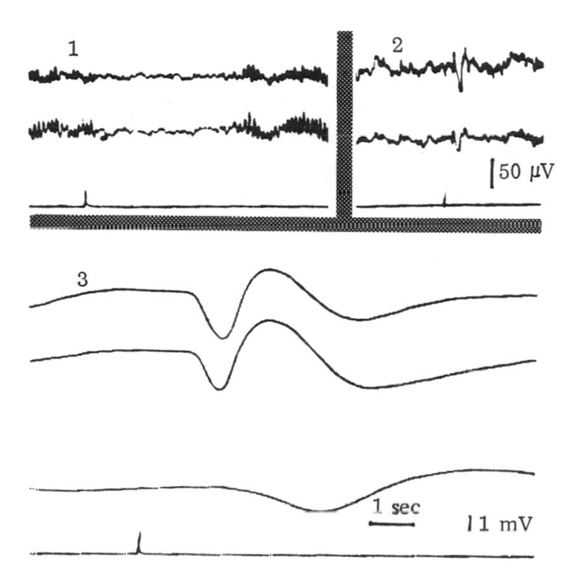

Fig. 2. Examples of the central (EEG) and efferent (skin potential) components of the orienting reflex. 1) Phenomenon of "desynchronization" (Subject No. D21, Tong X., 22-year-old male, recorded August 29, 1962.) From top to bottom: right occipital, left occipital EEG's, flash. 2) "On response (Subject No. A206, Shu X., 20-year-old female, recorded April 20, 1961). From top to bottom: left parieto-temporal, right parieto-temporal EEG's, click. 3) Phasic GSR (Subject No. D38, Ding X., 25-year-old female, recorded Sept. 6, 1962). From top to bottom: right hand, left hand, left foot, single cutaneous stimulus.

such that at the same time the waveforms of the two were the inverse, or almost the inverse of one another [see Fig. 2(3)].

The conditions of adaptation of the efferent (skin potential) component of the orienting reflex are summarized in Table 5. In adults and young people, the GSR's evoked by light stimuli were uniformly easily extinguished; only for cutaneous stimuli was the orienting reflex more difficult to extinguish in young people than in adults. For children of under seven years, the results were different; not only was the GSR evoked by all types of sensory input relatively strong, but also its extinction was difficult (see Table 6).

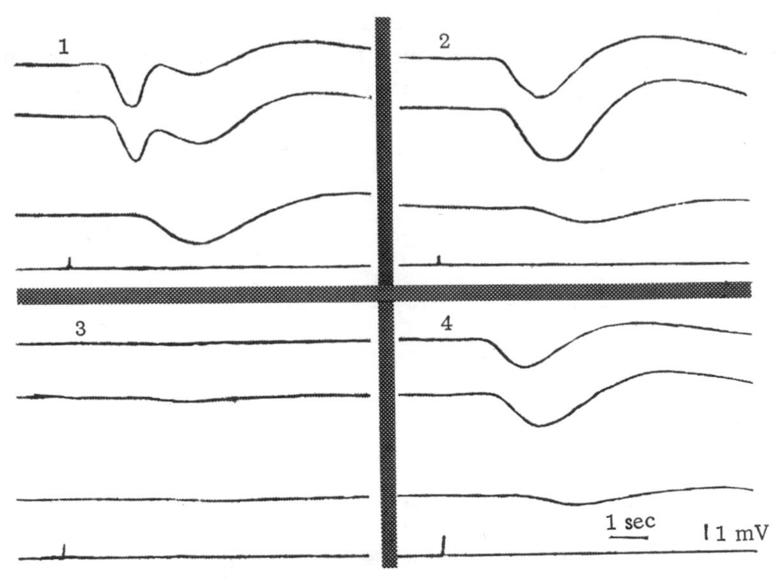

Fig. 3. Example of an unconditioned orienting reflex (Subject No. D40, Jou X., 32-year-old male, recorded Sept. 7, 1962). From top to bottom: electrodermograms of right hand, left hand, and right leg, respectively, and stimulus. 1) Fourth pairing of "cutan. +light"; 2) 18th pairing of "cutan. +light"; 3) after the 22nd pairing of "cutan. +light," the orienting reflex is extinguished; 4) a cutaneous stimulus alone evokes an unconditioned orienting reflex.

3. The Problem of the Unconditioned and the Conditioned Orienting Reflex

The orienting reflex can either be an unconditioned response or a conditioned response; this is entirely a matter of what the instantaneous neural mechanism of production of the response was at the particular moment. We clarified this question by carrying out further experiments on 26 subjects. In 20 subjects, a phasic GSR was evoked by single cutaneous and single light stimuli, and then the circumstances of the adaptation induced by paired "cutaneous + light" stimuli were examined. After the GSR evoked by "cutaneous + light" stimuli had been completely extinguished, the subjects were stimulated with single cutaneous stimuli, which again evoked a phasic GSR. We consider the latter to be an unconditioned orienting response, since with reinforcement of the adaptation, a "model" of the paired "cutaneous + light" stimulus is gradually formed by the subjects, and the destruction of the "model" acts as a novel agent, which in turn evokes an unconditioned orienting reflex (see Fig. 3). The original single cutaneous stimulus evokes a phasic GSR, the GSR evoked by the first "cutaneous + light" stimulus is biphasic, becoming monophasic after the 18th stimulus and completely disappearing after the 22nd, subsequent to which a single cutaneous stimulus again evokes a phasic GSR. In the remaining 6 subjects, we observed that single cutaneous stimuli evoked only "incomplete" desynchronization; single light stimuli, however, evoked a clear, "complete" desynchronization. Afterwards, we first obtained alpha-wave changes induced by cutaneous stimuli as an indifferent stimulus, then presented paired "cutaneous + light" stimuli to the subjects for 4 to 7 times, whereupon the presentation of single cutaneous stimuli then evoked clear, "complete" desynchronization. We consider, then, this "complete" desynchronization to be a conditioned orienting response, since the "cutaneous + light" combination resulted in the establishment of a temporary connection between them, as a result of which a cutaneous stimulus alone then evoked a conditioned orienting reflex (see Fig. 4).

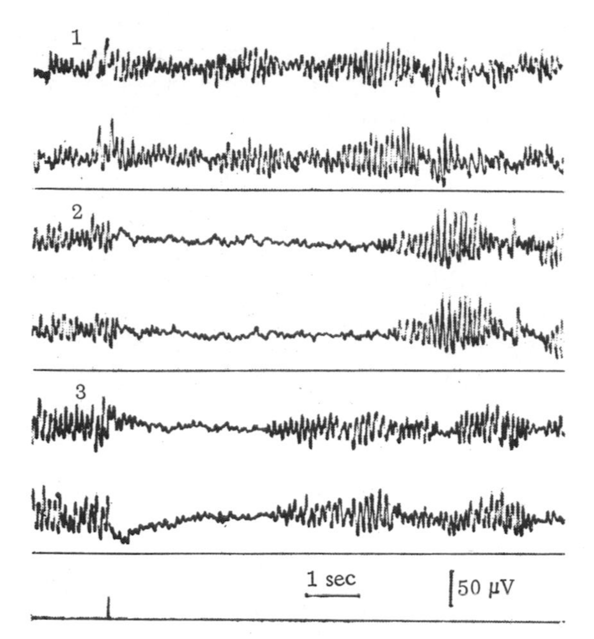

Fig. 4. Example of a conditioned orienting reflex (Subject No. D31, Hui X., 29-year-old female, recorded October 20, 1961). From top to bottom: right occipital, left occipital EEG's, stimuli. 1) A single cutaneous stimulus evokes "incomplete" desynchronization; 2) 5th pairing of "cutan. +light"; 3) cutaneous stimulus alone evokes a conditioned orienting reflex.

DISCUSSION

The orienting reflex is an adaptative response. When a novel agent appears in the environment, sensory receptors are then activated, there is orientation toward the direction of the novel agent, and at the same time the body is mobilized to respond to it. Any agent can be a signal, but if its informational content is unusually uncertain, this determines whether the signal has appeared once or repeatedly, for many times. From the standpoint of analysis of behavior, the basis of the orienting reflex is at the present time very actively discussed internationally. The question of how organisms perceive the surrounding objects and how they respond to these objects, how experience is reflected in the final output, and the related questions of attention and its counterpart, adaptation, etc., are of very considerable interest. Evidently, learning is related to the inhibition or blocking of unsuitable pathways, for example, the capability of forming an integrated response is dependent on the extinction of a large number of unsuitable responses initially. Galambos [16] pointed out that in the process of establishment of defensive reflexes in the monkey, the earliest signals evoke piloerection, a cry, and other types of "emotional" activity, but when the 50% level of correctness is attained in the course of learning, the signal evokes an "attention" response, and after the learning has become complete, the signal evokes only the specific effect or response. Consequently, in the investigation of the neural basis of learning, the understanding of the basic mechanism of adaptation is of the greatest importance.

Adaptation is a physiological phenomenon, being present in almost all aspects of the living activity of organisms. In carrying out research on higher animals and in man, it is first of all necessary to differentiate those processes arising peripherally from ones arising centrally; the orienting reflex can then be understood as a very good indicator of the latter. The representation for study of all complex things is the same: in order for information to have a relatively high entropy and a relatively small redundancy, the simplest or most fundamental "model" must effectively be found. Our data make clear that inputs of patterns of single stimuli are convenient for examining the simplest and most fundamental "model" of the orienting reflex. Utilizing this "model" permits study of the characteristics of the orienting reflex evoked in adults and children by a single sensory input. We believe the results of the above-mentioned investigation are fundamental to the investigation of the complex orienting response.

We now consider the question of the sensitivity of the EEG and GSR as indicators of the orienting reflex. It is well known that the phenomenon of alpha inhibition is an indicator of the orienting reflex (Gershuni [6]; John [19]); the GSR and mental activity are closely related (Wechsler [26]); Kuno [21] and others have termed the palmar areas of the hand and foot "psychic perspiratory areas." At the present time, many workers agree that the GSR and the orienting reflex are closely related (Mundy-Castle and McKiever [24]; and others). Sokolov [12] stated that the central (EEG) component of the orienting reflex possesses all of the characteristics of the efferent component, but our data from investigation of the orienting reflex from the developmental point of view show that there are prominent differences between the central (EEG) and efferent (skin potential) components, and for this reason, their neural mechanisms may be different. If we divide the GSR changes into tonic (slow spontaneous change) and phasic types, then in children of under seven years of age, the tonic change of the efferent (skin potential) component is relatively prominent, and the phasic change is relatively difficult to extinguish, but the central (EEG) component is extinguished comparatively rapidly. For adults, however, tonic change of the efferent (skin potential) component is not prominent, and the phasic change is easily extinguished, whereas the central (EEG) component is relatively difficult to extinguish. We believe that these differences are the products of two different neural structures in different locations. We agree with the view of some workers (Anokhin [2]; Jasper [18]; Lindsley [22]; and others) that the phenomenon of alpha-wave inhibition as a component within the overall orienting reflex is principally a function of subcortical structures, and only in this way can the orienting response be capable of being generalized, diffuse, and nonspecific in nature. For example, the phenomenon of alpha inhibition is frequently a product of the excitation of the subcortical brain stem reticular formation. More precisely, the central (EEG) component of the orienting reflex is in actuality a product of the mutual effects of cortical and subcortical structures, but the efferent (skin potential) component is perhaps principally a product of the subcortical reticular formation. The cells of the cerebral cortex of the pre-school child are not yet completely matured, and adaptability is lacking; for this reason, the central (EEG) component is incapable of being a very sensitive reflection of subcortical excitation.

In relation to the question of whether the orienting reflex is an unconditioned or a conditioned reflex, we believe that it can be both. The orienting reflex is an unconditioned reflex with respect to its occurrence; it regulates the activity of analyzers, so as to insure the best conditions for the organism's reception of external stimuli. The course of the orienting reflex is, however, different from an unconditioned reflex in that it is easily extinguished. From this point of view it is evident that the orienting reflex perhaps more nearly resembles a conditioned reflex rather than an unconditioned one. However, the orienting reflex is different from a conditioned reflex in that a novel stimulus frequently results in the inhibition of a conditioned reflex, whereas such a stimulus results in the excitation of the orienting reflex. In the process of establishing a conditioned reflex, the activity of the specific system is initially weak, whereas the activity of an orienting reflex is strong. Contrariwise, after reinforcement of a conditioned reflex, the activity of the orienting reflex then becomes weak, whereas the activity of the specific system is stronger. We believe that the same phenomenon of alpha inhibition or of a phasic skin potential change can be not only an unconditioned response but also a conditioned response, the key point of the question being the neural mechanism at the moment of appearance of these phenomena. For instance, our experimental results attest that the first example is an unconditioned response and that the second example is a conditioned one. Although the skin potential was used as an indicator in the first example, and the EEG was used in the second, this does not in the least indicate that unconditioned or conditioned responses are necessarily dependent on the particular indicator. For example, in our paper, "Electroencephalographic study of temporary connections in man" [1], the skin potential was used as an indicator for the establishment of a conditioned response. It is of interest that we were able to establish it using the orienting reflex as basically a conditioned reflex. This type of conditioned reflex differs from the usual conditioned reflex in that it is extinguished especially rapidly, which possibly results for the reason that in using the orienting reflex as the basis of a conditioned reflex, the characteristics of the unconditioned reflex must be "reproduced."

CONCLUSIONS

We used different types of single sensory inputs (weak light, sound, and cutaneous stimuli) and the EEG and galvanic skin potential (electrodermogram) to investigate the "neural model" of the orienting reflex. The experimental results were as follows:

1. Single stimuli were found to be suitable inputs for study of the "model" of the orienting reflex; the characteristics of the responses and of their extinction in the manifestation of the "model" were different for different types of single sensory inputs.

2. The behavior of the central (EEG) and efferent (skin potential) components of the orienting reflex did not completely parallel one another. For example, in children of under 7 years of age, the former was relatively weak whereas the latter was comparatively prominent; in adults, however, the former was prominent whereas the latter was comparatively weak.

3. The orienting reflex can be both an unconditioned and a conditioned response.

REFERENCES

1. Liu Shih-yih and Wu Qin-e, [Electroencephalographic study of temporary connections in man] (in Chinese), Acta Psychol. Sinica, No. 1, pp. 11-19 (1963) (See pp. 75-85, this volume).
2. Anokhin, P. K., [The role of the orientational-investigatory response in the formation of the conditioned reflex, in: The Orienting Reflex and Orientation-Investigatory Activity] (in Russian] (1958), pp. 9-20.
3. Asratian, E. A., [On the physiology of temporary connections, in: Fifty Years of the Teaching of Pavlov on the Conditioned Reflex] (in Russian), Moscow (1952), pp. 68-99.
4. Asratian, E. A., [Reply to the criticism of A. G. Ivanov-Smolenskii] (in Russian), Zh. Vyssh. Nerv. Deiat., 3(4):636-648 (1953).
5. Voronin, L. G., and Sokolov, E. N., [On the question of the mechanism of the orienting reflex and its interrelation with the conditioned reflex, in: Mechanisms of the Development of Temporary Connections, Gagra Conferences, Vol. III] (in Russian), ed. I. S. Beritashvili (Beritov), Publishing House of the Academy of Sciences of the Georgian SSR, Tbilisi (1960), pp. 213-227. (To be published in English translation by the American Psychological Association.)
6. Gershuni, G. V., [Reflex responses to the influence of external stimuli on sensory organs of man and their relation to sensation] (in Russian), Fiziol. Zh. SSSR, 35(5):549-560 (1949).
7. Narbutovich, I. O., and Podkopaev, N. A., [The conditioned reflex as an association] (in Russian), Trud. Fiziol. Lab. I. P. Pavlova, 6(2):5-25 (1936).
8. Novikova, L. A., and Sokolov, E. N., [Investigation of the electroencephalogram of motor and galvanic skin responses in orienting and conditioned reflexes in man] (in Russian), Zh. Vyssh. Nerv. Deiat., 7(3):363-372 (1957).
9. Pavlov, I. P., ["Internal inhibition" of conditioned reflexes and sleep—one and the same process] (in Russian), Complete Works, Vol. 3 (1949), pp. 296-309.
10. Podkopaev, N. A., [The conditioned reflex as an association] (in Russian), Fifth Congress of Physiologists (1934), p. 62.
11. Sokolov, E. N., [The orienting reflex, its structure and mechanisms, in: The Orienting Reflex and Orientational-Investigatory Activity] (in Russian) (1958), pp. 111-122.
12. Sokolov, E. N., [On the question of the galvanic skin component of the orienting reflex, in: The Orienting Reflex and Problems of Higher Nervous Activity] (in Russian) (1959), pp. 52-76.
13. Sokolov, E. N., [The orienting reflex, in: The Orienting Reflex and Problems of Higher Nervous Activity] (in Russian) (1959), pp. 5-51.
14. Sokolov, E. N., and Paramonova, N. P., [Dynamics of the orienting reflex in the development of sleep inhibition in man] (in Russian), Zh. Vyss. Nerv. Deiat., 11(2):206-215 (1961).

15. Sokolov, E. N., and Paramonova, N. P., [On the extinction of the orienting reflex] (in Russian), Zh. Vyss. Nerv. Deiat., 11(1):3-11 (1961).
16. Galambos, R., Changing concepts of the learning mechanism, Brain Mechanisms and Learning Symposium (1961), pp. 231-241.
17. Grastyân, E., The significance of the earliest manifestation of conditioning in the mechanism of learning, Brain Mechanisms and Learning Symposium (1961), pp. 243-263.
18. Jasper, H., Reticular-cortical systems and theories of the integrative actions of the brain, Biological and Biochemical Basis of Behavior (1958), pp. 37-61.
19. John, E. R., Higher nervous functions: brain functions and learning, Ann. Rev. Physiol., 23:451-484 (1961).
20. Jouvet, M., Recherches sur les mēchanismes neurophysiologiques du sommeil et de l'apprentissage nēgatif, Brain Mechanisms and Learning Symposium (1961), pp. 445-479.
21. Kuno, Y., Human Perspiration (1956).
22. Lindsley, D. B., Psychophysiology and perception, in: Current Trends in the Description and Analysis of Behavior (1958), pp. 48-91.
23. Motokawa, K., Electroencephalograms of man in the generalization and differentiation of conditioned reflexes, Tohoku J. exp. Med., 50:225-234 (1949).
24. Mundy-Castle, A. C., and McKiever, B. L., The psychological significance of the galvanic skin response, J. exp. Psychol., 46(1):15-24 (1953).
25. Segundo, J. P., Roig, J. A., and Sommer-Smith, J. A., Conditioning of reticular formation stimulation effects, Electroenceph. clin. Neurophysiol., 11:471-484 (1959).
26. Wechsler, D., The Measurement of Emotional Reactions (Researches on the Psychogalvanic Reflex) (1925).
27. Voronin, L. G., and Sokolov, E. N., Cortical mechanisms of the orienting reflex and its relation to the conditioned reflex, Electroenceph. clin. Neurophysiol. Suppl., 13:335-346 (1960).

ELECTROENCEPHALOGRAPHIC STUDY OF TEMPORARY CONNECTIONS IN MAN*

Liu Shih-yih and Wu Qin-e†

The electroencephalographic study of the mechanism of formation of temporary connections has received special attention in the last ten years, but the great majority of studies have been concerned with the EEG in animals. These studies cannot but alter some recognized basic concepts of the mechanisms of formation of classical conditioned reflexes. For example, we now know that the formation of conditioned reflexes is not only a mechanism of the cerebral cortex, but one that involves the entire brain from the reticular formation to the cerebral cortex. At the present time, it is generally recognized that the site of the very first "encounter" of the inputs (the conditioned and unconditioned stimuli) is in the brain stem reticular formation; in the process of progressive formation of a conditioned reflex, activity of the neurons of the brain stem reticular formation gradually comes to resemble the activity of the cells of the thalamic reticular formation, and the connections between thalamic and cortical cells are considered to form the basis for the reinforcement and formation of conditioned reflexes. A relatively detailed electroencephalographic study of temporary connections in man has been carried out by Gastaut et al. [12, 13] and studies have been carried out by a number of other workers, including Jasper and Shagass [15], Motokawa [23], Kozhevnikov [5], Stevens [25], and Kratin [6], but the large majority of investigators have described the formation of a particular reflex. We believe it is important that several kinds of indicators be selected to investigate the characteristics and neurophysiological mechanisms of the formation of temporary connections in the human brain.

Traditionally, the phenomenon of the inhibition of alpha activity by light stimuli as the indicator of the unconditioned response has been used in the electroencephalographic study of temporary connections, the conditional stimulus being a weak sound or a mild touch that does not evoke inhibition of the alpha. After some number of pairings with the light, the latter stimulus alone can evoke an inhibition of the alpha, which indicates the formation of a conditioned reflex (Knott and Henry [18], Loomis et al. [19], Morrell and Ross [22], Gastaut [13], Wells [29], and Peimer [9]). In practice, in this type of study it is convenient to take the orienting reflex as the basis for the establishment of a "sensori-sensory" conditioned response. But the input for this type of conditioned reflex is generally a 1-sec continuous or interrupted stimulus pattern, and heretofore patterns of a single impulse stimulus have not been used to investigate this problem. From the standpoint of brain control theory, however, patterns consisting of single stimuli are of very great help in understanding the problem of the so-called "neural

---

*Acta Psychologica Sinica, No. 1, pp. 11-19 (1963).

†Cheng Su-zhen participated in this work.

model" advanced by Voronin and Sokolov [27]. Moreover, the use of characteristics of the orienting reflex corresponding to the input to different analyzers for investigating the problem of birectional conditioned reflexes (i.e., having both EEG and efferent components) for patterns of single stimuli, which has also heretofore not received attention, is actually very helpful in understanding the problem of the mechanism of formation of temporary connections.

Whether the "assimilation of rhythm," as the basis of a new type of formation of conditioned reflexes, can be taken directly for the investigation of temporary connections in man is an unresolved question. Yoshii et al. [31] considered the "assimilation of rhythm" type of conditioned reflex as the beginning of a new era in the electroencephalographic investigation of temporary connections, that is, the frequency of stimulation can become a conditioned reflex and be reflected in the EEG. Magoun [20] pointed out that investigation of the "assimilation of rhythm" type of conditioned reflex is like employing a simple tracer to follow the processes that occur in the brain. Work in this respect began in 1955 (Morrell and Jasper [21]), and afterwards the studies of many workers (Yoshii and Hockaday [30], John and Killam [17], Ulett et al. [26]), affirmed this type of temporary connection in the animal brain. But several workers (for example, Gastaut et al. [13]) pointed out that in the brain of human subjects who clearly showed the phenomenon of "assimilation of rhythm," it was uniformly impossible to use the frequency of the light to establish conditioned reflexes of the "assimilation of rhythm" and alpha inhibition types, respectively. Stevens [25] recently mentioned that Grey Walter, using implanted electrodes, was also unable to establish an "assimilation of rhythm" type of conditioned reflex. Other workers (Jus and Jus [16], and Bercel et al. [11]) were inclined to an affirmative view on this question, but thus far we have not seen reports of experimental studies concerning the formation of the above-mentioned conditioned reflexes in the human brain.

The purpose of the present investigation was to examine, with the aid of new indicators, three basic theoretical problems concerning the electroencephalographic study of temporary connections in the human brain: (1) the problem of establishing "sensori-sensory" conditioned responses by patterns of single impulse stimuli, (2) the problem of establishing "sensori-sensory" bidirectional conditioned responses by pulsed single stimuli, and (3) the problem of establishing the "assimilation of rhythm" type of conditioned responses in the human brain.

METHODS

The subjects were 10- to 45-year-old normal children and adults (including primary, intermediate, and high school students and members of organizations). The number of subjects in the first, second, and third series of experiments were 102, 59, and 80, respectively. In the first series of experiments, the EEG, primarily, was recorded; in the second and third series of experiments, the EEG and GSR, and in some subjects also the EKG, were recorded simultaneously. The instrument used for the recordings was a 16-channel Ediswan-III ink-writing electroencephalograph. The EEG's were principally bipolar occipital recordings; at times temporal, parietal, frontal, and fronto-parietal potentials were also recorded, an ear being used for the indifferent electrode and as a ground. For the recordings of skin potentials, 2-cm silver disc electrodes were applied with the aid of special electrode paste to the dorsal and palmar surfaces of the left hand. For the EKG recordings, silver electrodes were affixed to the left and right wrists. The light and sound stimuli were generated from a "pulse generator" and "electrolight" and "electroacoustic" transducers of our own manufacture. The "pulse generator" controlled the interval (frequency), duration (width), and amplitude of single or repetitive brief pulses, for simultaneously triggering the "electrolight" and "electroacoustic" transducers. The stimuli chosen for the present study were relatively low-amplitude light (25 lux) and sound (25 db) pulses. For the cutaneous stimuli, the customary Pavlovian laboratory skin stimulating apparatus was used and affixed to the medial aspect of the left leg of the subjects. The pulses for the light and sound pulses were electrically transmitted, being linked by the same delay to the $^{1}/_{15}$-sec brief trigger pulse; consequently, in the first type of experiment, the stimuli for single "sound + light" combinations were almost completely simultaneous. The cutaneous stimuli, however, were transmitted by an air system; consequently, for combinations of "light + cutaneous" stimuli, the stimuli were not completely simultaneous, i.e., the cutaneous stimuli were slightly delayed with respect to the light stimuli. During the experiments, the subject sat on a sofa in a specially constructed soundproof, electrically shielded, darkened room. The experimenter maintained contact and transmitted instructions to the subjects by means of a two-way intercom system.

In the first series of experiments, single sound and light stimuli, respectively, were initially used, to observe which subjects showed the phenomenon of "desynchronization." For subjects for whom single light stimuli clearly evoked "desynchronization" and for whom single sound stimuli rapidly become indifferent stimuli, we then carried out a study of "sensori-sensory" conditioned reflexes to single "sound + light" stimuli. In the second series of experiments, we initially used single light and sound stimuli, respectively, to observe the subjects' "desynchronization" and the electrical skin responsivity. If subjects clearly showed "desynchronization" in response to single light stimuli, but the latter stimuli rapidly became indifferent for skin electrical responses, and further, if there were clear cutaneous electrical responses to single cutaneous stimuli, but the latter stimuli rapidly became indifferent with respect to alpha waves, we then carried out an investigation of bidirectional "light + cutaneous" conditioned reflexes. In the first and second experimental series, many repetitions of 10 "sensori-sensory" pairings were carried out on each experimental day, the time interval between each successive pairing varying between $\frac{1}{2}$ to 2 minutes. In the third series of experiments, we carried out an investigation of conditioned reflexes of a "cutaneous rhythm — light rhythm" type in those children and young people (10 to 18 years old) who clearly showed the phenomenon of the "assimilation of rhythm," but for whom cutaneous stimuli rapidly became indifferent with respect to alpha waves. Initially, a 1.5/sec to 2/sec skin stimulus was applied alone for 1 to 2 sec, and then combined with a rhythmic light stimulus for 6 to 7 sec. The frequency of the flicker chosen for each subject was that corresponding to an optimum "assimilation of rhythm." On each experimental day, many repetitions of approximately 10 pairings were carried out, the intervals between successive pairings being randomized from 1 to 3 minutes.

RESULTS

1. Establishment of "Sensori-sensory" Conditioned Reflexes by Patterns of Pulsed Single Stimuli

We used evoked potentials to investigate the effect of the input (stimulation of sensory organs or of sensory pathways) on the central nervous system. Evoked potentials can be divided into single and repetitive types. The important data from the EEG of animals have almost all come from electrical responses evoked by patterns of single stimuli, but there is very little data from evoked potentials to single stimuli in the human EEG. It was necessary for us to examine whether it is possible to use patterns of pulse-type single stimuli to establish "sensori-sensory" conditioned reflexes in the human brain. The experiments were carried out in the soundproof, electrically shielded darkened room, employing pulses of weak sound and weak light of 0.5 sec maximum duration. The single stimuli were triggered from the same trigger pulse; consequently, the "sensori-sensory" pairings were almost completely simultaneous. We observed the spontaneous EEG potentials of 102 young subjects, and then, using light and sound pulses, observed their EEG's for "desynchronization"; from the results, 54 subjects from among them were eliminated, because the spontaneous potentials of these subjects were not prominent or the sound pulses did not rapidly become indifferent stimuli. The investigation of "sound + light" conditioned reflexes was carried out with the remaining 48 subjects. The stimuli

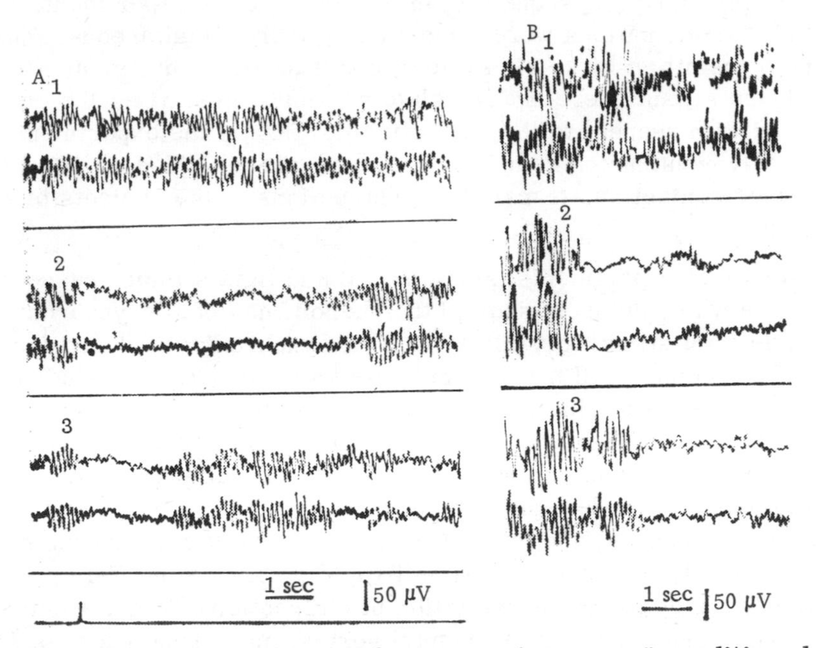

Fig. 1. Example of the formation of a "sensori-sensory" conditioned reflex to pulsed single stimuli. From top to bottom: right occipital EEG, left occipital EEG, and stimulus. A. Weak "desynchronization" evoked by weak sound rapidly becomes extinguished, a weak light evokes "complete" desynchronization; after 3 pairings of "sound + light," a single weak sound evokes "complete" desynchronization. (Subject No. Y2, Yi X., 16-year-old female, recorded April 20, 1960.) 1) Weak sound stimulus has already become indifferent with respect to alpha waves. 2) Third pairing of "sound + light." 3) Single weak sound evokes a "complete" desynchronization. B. Weak sound evokes an "incomplete" desynchronization, weak light evokes "complete" desynchronization; after 5 pairings of "sound + light," a single sound evokes "complete" desynchronization. (Subject No. C1276, Ai X., 10-year-old female, recorded May 22, 1961.) 1) Weak sound evokes an "incomplete" desynchronization. 2) Fifth "sound + light" pairing. 3) Single weak sound evokes a "complete" desynchronization.

were generally paired for 4 to 7 times, after which the recordings were inspected. The results showed that "sensori-sensory" temporary connections for patterns of single stimuli were formed in 29.5% of subjects (14 persons) (see Fig. 1). In subject No. Y2 (Fig. 1A), "desynchronization" was evoked by the first weak sound only, and the stimuli after the second one rapidly became indifferent. Then, after the third "sound + light" pairing, single sound evoked the phenomenon of "desynchronization." For subject No. C1276 (Fig. 1B), weak sound stimuli evoked only "incomplete" desynchronization, but light stimuli evoked "complete" desynchronization. After the fifth "sound + light" pairing, single sound stimuli evoked "complete" desynchronization. It should be pointed out that the formation of this type of "sensori-sensory" temporary connection is unstable and very easily extinguished. Generally, after 5 to 7 "sound + light" pairings, at most two repetitions of the sound stimuli alone will evoke clear conditioned responses, after which the response gradually disappears. If the number of repetitions of the pairings is sufficiently great, or is continued after the appearance of the conditioned response, then the conditioned response rapidly weakens or the sound stimulus gradually becomes an indifferent stimulus.

2. Formation of Bidirectional Conditioned Reflexes by Patterns of Single Stimuli

In a study of the EEG and GSR in the orienting reflex in man [3], we pointed out that the components of the orienting reflexes evoked by dissimilar single stimuli exhibited selectivity, and we utilized this characteristic to examine the question of bidirectional conditioned reflexes formed in the human brain by single stimuli. We first used light and cutaneous stimuli to characterize the "desynchronization" and skin responses in 59 subjects, with the result that 13 subjects were eliminated because of their spontaneous potentials, or because their "desynchronization" to light was not prominent. For the 46 remaining subjects, the duration of the latter for later single light stimuli was shorter than that for the earlier ones, but the "desynchronization" phenomenon still remained intact. Single stimuli could evoke phasic cutaneous electrical responses in a part of the subjects, although monophasic positive waves were frequently seen. After repetition (generally 2 or 3 times), single light stimuli rapidly became indifferent for the galvanic skin response. Among these 46 subjects, we did not find any for whom a single cutaneous stimulation of the skin evoked a skin response, and there were only 25 subjects in whom a prominent phasic skin response could be evoked. The latter was frequently biphasic, consisting of an initial comparatively short positive potential, followed by a comparatively long negative potential. Following the latter, another positive potential could sometimes also be seen. Although with repetition of the stimulus there was a simplification of waveform, decrease in the amplitude of the waves and an increase in the latent period, the phasic change could still be clearly seen. Single cutaneous stimuli could evoke the phenomenon of "desynchronization" in most subjects, but after two or three repetitions, the stimulus had already become indifferent with respect to alpha waves.

In view of the above-mentioned findings, we were able to study bidirectional conditioned reflexes with single stimuli in only 25 of the 59 subjects. Thus, single light stimuli could evoke the central (EEG) component of the orienting reflex in these subjects, but such stimuli rapidly became indifferent or almost indifferent with respect to the efferent (skin potential) component of the orienting reflex, and single cutaneous stimuli could evoke the efferent (cutaneous potential) component of the orienting reflex, but this stimulus rapidly became indifferent with respect to the central (EEG) component of the orienting reflex. On these subjects, we carried out studies on "sensori-sensory" bidirectional conditioned responses of the "light + cutaneous" type. The "light + cutaneous" stimuli were generally paired for 4 to 7 times, after which trials with single light or cutaneous stimuli were carried out separately. The results showed that 24.4% of subjects (11 individuals) showed cutaneous responses to single light stimuli, and EEG "desynchronization" appeared in response to single cutaneous stimuli, i.e., bidirectional conditioned responses were formed. From Fig. 2, it is apparent that prior to the formation of the conditioned reflex, single light stimuli evoked only a monophasic positive potential of low amplitude that rapidly faded; single cutaneous stimuli also evoked a slight "desynchronization," which also rapidly diminished with repetition of the stimuli. After 6 pairings of single "light + cutaneous" stimuli, single light stimuli alone evoked a high-amplitude positive-negative biphasic response, and single cutaneous stimuli evoked clear alpha inhibition. This type of bidirectional temporary connection was also characterized by instability and ready extinction. If the "light + cutaneous" pairing were continued after the conditioned response appeared, then the conditioned response rapidly diminished, i.e., the stimuli gradually became indifferent.

3. Formation of the "Assimilation of Rhythm" Type of Conditioned Reflex in Man

Many workers believe that it is impossible to establish an "assimilation of rhythm" type of conditioned reflex in the human brain. Gastaut emphasized that even subjects who showed

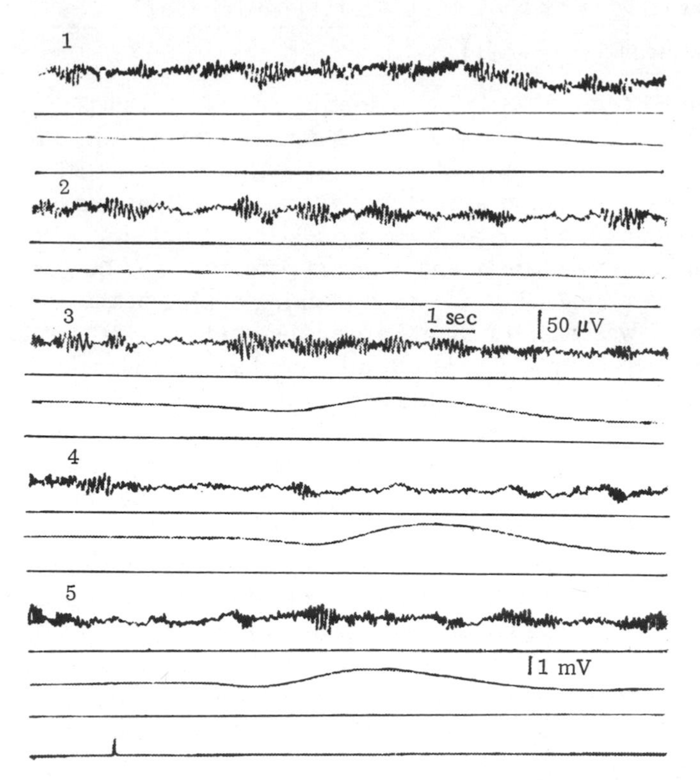

Fig. 2. Example of the formation of bidirectional conditioned reflexes by single stimuli. (Subject No. D30, Tang X., 31-year-old female, recorded October 20, 1961.) From top to bottom: right occipital EEG, left electrodermogram, and stimulus. 1) The "desynchronization" resulting from a single cutaneous stimulus has already disappeared. 2) The phasic skin potential response to single light stimuli has already become extinguished. 3) Sixth pairing of "cutaneous + light" stimuli. 4) Single cutaneous stimulus evokes "desynchronization." 5) Single light stimulus evokes a phasic skin potential response.

"assimilation of rhythm" prominently could not form "desynchronization" and "assimilation of rhythm" types of conditioned reflexes. Our previous study [2] showed that in 12- to 16-year-old subjects it is very easy to obtain very distinct, and as a rule persistent, "assimilation of rhythm"; it was for this reason that 80 subjects of approximately this age level were selected for the present study. But 14 of these subjects were subsequently eliminated because at the time of the second recording, their "assimilation of rhythm" was not sufficiently prominent or was smaller than for the first recording. At the beginning of an experiment, the most suitable frequency for evoking the "assimilation of rhythm" in each subject was sought, for use subsequently as the conditioning stimulus. For most subjects, frequencies of 4/sec to 7/sec were the most suitable.

Before beginning the establishment of a conditioned reflex, an interval of 7 to 9 sec of rhythmic cutaneous stimulation (at a frequency of 1.5/sec to 2/sec) was carried out to effect extinguishment, so that this stimulus would become indifferent with respect to alpha waves. The rhythmic weak cutaneous stimuli could not cause the occipital alpha to produce the phenomenon of "assimilation of rhythm"; they could only evoke very weak "desynchronization," the "desynchronization" appearing only for the initial few seconds of stimulation, generally requiring only 3 to 4 repetitions before rapidly becoming extinguished. Following this, "sensori-sensory" conditioned reflexes were formed with "rhythmic cutaneous + rhythmic light" stimuli. Rhythmic

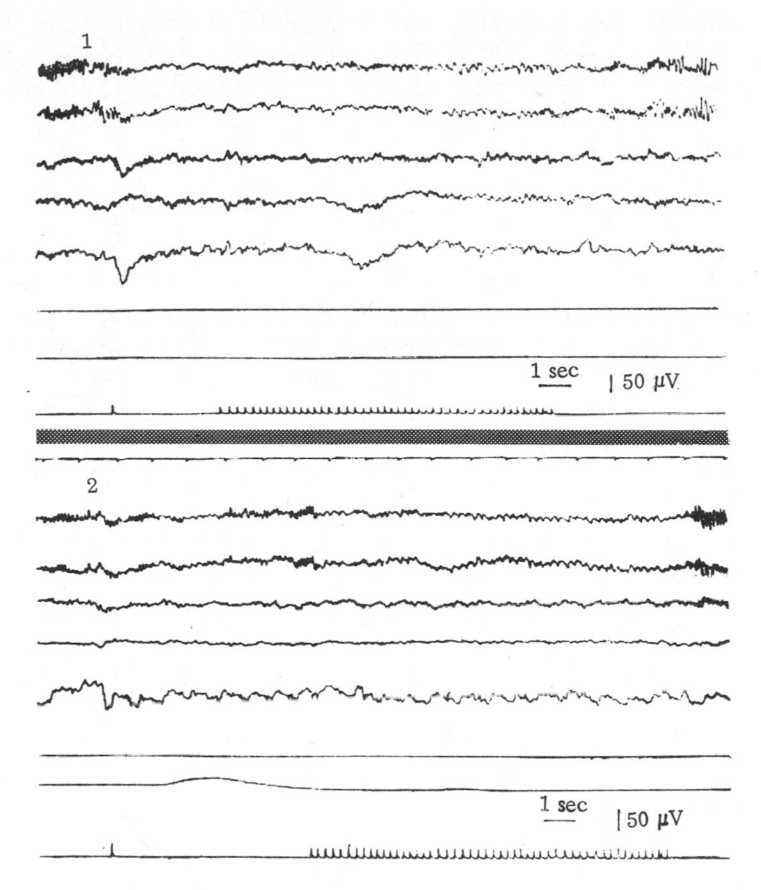

Fig. 3. Example of the formation of "desynchronization" and "assimilation of rhythm" types of conditioned reflexes in young subjects who showed the phenomenon of "assimilation of rhythm." 1) Example of the formation of a "desynchronization" type of conditioned reflex in a child who showed the phenomenon of "assimilation of rhythm." (Subject No. D61, Qi X., 10-year-old female, recorded June 25, 1962.) From top to bottom: right occipital, left occipital, right parieto-temporal, frontal, left fronto-parietal EEG's, blank channel, left hand skin potential, stimulus. In the sixth pairing "cutaneous + rhythmic light," a conditioned response of "desynchronization" appears. 2) Illustrative example of the establishment of an "assimilation of rhythm" type of conditioned reflex in a subject showing the phenomenon of "assimilation of rhythm." (Subject No. D42, Lü X., 17-year-old male, recorded June 8, 1962.) From top to bottom: time marker (1 sec), right occipital, left occipital, right parieto-temporal, frontal, left fronto-parietal, blank channel, skin potential of the left hand, stimulus. An "assimilation of rhythm" type of conditioned response appears in the interval after the cutaneous stimulus of the 12th pairing of "cutaneous + rhythmic light" pairing.

cutaneous stimuli were first applied for 1 to 2 sec, then combined with rhythmic light stimuli, for 6 to 7 sec. The occipital alpha waves were observed for any changes after the initiation of cutaneous stimuli but before combination with the rhythmic light. This "testing" method consisted of presenting the cutaneous stimuli for 2 to 3 sec, then adding the rhythmic light, and observing the latency for the resulting EEG changes. This "test" was repeated at least 3 to 5 times.

Our results showed that the formation of a "rhythmic cutaneous + rhythmic light" type of "sensori-sensory" conditioned reflex can have two forms: one type being a "desynchronization" [Fig. 3(1)], the other being an "assimilation of rhythm" [Fig. 3(2)]. The formation of these two types of conditioned reflexes could have independent neuronal mechanisms, but from many data derived from the investigation of the EEG of animals it is apparent that they could represent two different forms of the same conditioned reflex. Our data showed that in 53% of subjects (35 individuals) the "desynchronization" type of conditioned reflex could be formed, and in 19.7% of subjects (13 individuals) an initial "desynchronization" followed by an "assimilation of rhythm" type of conditioned reflex could be formed. This indicates that only for those subjects who are able to form the "desynchronization" type of conditioned reflex can the "assimilation of rhythm" type of conditioned reflex be formed thereafter. On the other hand, however, most subjects who were able to form a "desynchronization" type of conditioned reflex could not subsequently form the "assimilation of rhythm" type of conditioned reflex. Irrespective of whether the "desynchronization" or the "assimilation of rhythm" types of conditioned reflexes were established, they were not at all stable and were unpredictable. The "assimilation of rhythm" types of conditioned reflexes formed were temporary and unstable; they were often replaced by "desynchronization" types of conditioned reflexes. For instance, taking subject No. 64 as an example, an "assimilation of rhythm" type of conditioned reflex appeared with the fourth and sixth pairing, but for the third, fifth, and twelfth pairing, only the "desynchronization" type of conditioned reflex appeared. Like the results from the brain potentials of animals, formation of the so-called "assimilation of rhythm" type of conditioned reflex is not as evident in response to cutaneous stimuli alone having a frequency equal to that for the appearance of a distinct assimilation of rhythm when "rhythmic cutaneous + rhythmic light" stimuli are used, but some assimilation does appear, the rhythm of which approximates that of the assimilation of the frequency of the rhythmic light. It is especially noteworthy that we observed in 4 subjects, in the course of the formation of an "assimilation of rhythm" conditioned reflex to "rhythmic cutaneous and rhythmic light" stimuli, that the spontaneous occipital electrical potentials also showed the phenomenon of "assimilation of rhythm" intermittently. Consequently, it is not possible that the previous information is produced directly from cortical cells, because its appearance then necessarily includes the activity of nerve cells of so-called "reverberating circuits," and necessarily includes "storage" or "memory."

DISCUSSION

The formation of conditioned reflexes by single stimuli is a new problem in the study of temporary connections. We have pointed out that pulsed single stimuli can be used for examining the "neuronal model" of the orienting reflex, and they are superior to continuous or intermittent stimuli [3]. Since they simplified the investigation, they had the result that the information had a comparatively high entropy and a relatively small redundancy. From the standpoint of brain control mechanisms, single stimuli are also very simple for investigating "sensori-sensory" temporary connections (although they cannot be used for examining conditioned reflexes generally). Classical studies long ago demonstrated the necessity of paying attention to the time relationships of conditional and unconditional stimuli, since only under the conditions that the conditional stimulus acts prior to the unconditional stimulus does the meaning of the signal correspond to the conditioned reflex. This rule, however, comes mainly from research on food reward conditioned reflexes (the so-called "conditioned-unconditioned" temporary connections described by Ivanov-Smolenskii [4]) and the conditions for "sensori-sensory" conditioned reflexes (conditioned-conditioned temporary connections described by Ivanov-Smolenskii) are different. Since most of the temporary connections we make in everyday life are of the "sensori-sensory" type, it is impossible to imagine that in each of the so-called "associations" established in everyday life the relations of their time order are necessarily fixed. It is noteworthy that almost completely synchronized pulses of single "sound + light" stimuli formed "sensori-sensory" temporary connections in the human brain; we consider this a very simplified but fundamental "model" for examining "sensori-sensory" conditioned reflexes. From the results of our study it is apparent that this "model" already shows at least the following two important facts: (1) during the formation of "sensori-sensory" conditioned reflexes, under conditions of two simultaneous conditional stimuli, the meaning of the signal of the conditioned reflex is the same; (2) the human brain can simultaneously receive information transmitted by two different pathways (but not from the same pathway) (Liu Shih-yih [1]).

The characteristics of orienting reflexes evoked by single stimuli from different analyzers are different. On the basis of this feature, we formed different components of the orienting reflex as the basis of a bidirectional conditioned reflex. Our experimental results showed, on the one hand, that when temporary connections are formed for two indifferent stimuli, it is necessary that each indifferent stimulus be able to evoke an orienting reflex (Narbutovich et al. [8], Podkopaev [10], Liu Shih-yih [7], etc.), but it is not necessary that each neutral stimulus be able to evoke all the components of a complete orienting reflex, or evoke the same components of the orienting reflex. On the other hand, the results show that the information transmitted by the area of representation in the central nervous system of two indifferent stimuli is bidirectional. The formation of the above-mentioned bidirectional conditioned reflex is fundamentally a "sensori-sensory" conditioned reflex, although the basic mechanisms of the central (EEG) component of the orienting reflex and the efferent (skin potential) component can differ. If we agree that the mechanism of the former is a mutual product of cortical and subcortical cells, then the mechanism of the closure of the latter must be different. Wang et al., in 1956 [28], showed that the galvanic skin response is closely related to the reticular formation; Sourek and Skorpil, in 1961 [24], using a neurosurgical operation, also demonstrated that the galvanic skin reflex was related to subcortical structures. Grastyán et al., in 1956 [14], considered that artificial orienting reflexes were very easy to evoke by stimulation of chronically implanted electrodes in the hypothalamus and reticular formation. Consequently, we consider that the basic mechanism of closure of the efferent (skin potential) component of the orienting reflex as a conditioned reflex is mainly related to the subcortical reticular formation.

Our data further demonstrate that subjects in the 10- to 18-year-old group having an especially prominent "assimilation of rhythm" are able to form "desynchronization" and "assimilation of rhythm" types of conditioned reflexes, but the percentages of the latter are relatively small, and they are not absolutely stable and are temporary. It is known that the formation of the "assimilation of rhythm" type of conditioned reflex in the brain of animals is generally in the following three phases: (1) widespread "desynchronization," (2) localized "assimilation of rhythm," and (3) localized "desynchronization" (Jus and Jus [16], Morrell and Jasper [21]). Our results also demonstrate that only in those subjects in whom the "desynchronization" conditioned reflex is evoked initially is it possible to evoke the "assimilation of rhythm" type of conditioned reflex subsequently. It is apparent that in the formation of the "assimilation of rhythm" type of conditioned reflex in the human brain, the second phase is unstable and easily extinguished. The reason for our particular interest in the phenomenon of "assimilation of rhythm" as a basic conditioned reflex is that the study of this type of conditioned reflex is helpful in disclosing the mechanisms of closure of temporary connections in the human brain, and it is also helpful in understanding the relationship between the afferent (present) information and retention (instantaneous recall).

In the third experimental part of our study, it was observed that "sensori-sensory" temporary connections formed in the human brain have the characteristic of being very easily extinguished. Temporary connections formed in the human brain are thus unlike the durable ones that are formed in animals. This is because "sensori-sensory" temporary connections are basic to thought and learning, and in the course of human life, a large number of temporary connections emerge every day, but not many of these persist, the majority being "instantaneous" only, and after a short time rapidly disappear in part or altogether. Some workers in the field of learning believe that there are directional changes that take place in the central nervous system which are determined synaptically. The random activity of any neuron can act as noise and interfere with learning, and there are at least several thousand times more neurons in the human brain than are necessary for learning at any given time, so that during learning at any time the number of remaining neurons is very great. Consequently, an important part of the process of the formation of temporary connections is the elimination of unsuitable ones, as well as the establishment of new ones. The reticular pathways of the human brain are very complex, and from the standpoint of analysis of adaptative processes, the "noise" level is of course higher than that of the animal brain; consequently, the very simple "sensori-sensory" temporary connections that are formed are very unstable.

CONCLUSIONS

1. We used the EEG as an indicator to demonstrate that "sensori-sensory" (sound + light) conditioned reflexes in man can be formed to patterns of pulsed single stimuli. These "sensori-sensory" pairings were almost completely simultaneous.

2. We used the EEG and the electrodermogram (skin potential) to demonstrate the formation of the central (EEG) component of the orienting reflex and the efferent (skin potential) components of bidirectional "sensori-sensory" conditioned reflexes in the human brain.

3. In young subjects showing the phenomenon of "assimilation of rhythm," it was possible to establish "desynchronization" and "assimilation of rhythm" types of conditioned reflexes, but the latter were not as prominent and were unstable.

REFERENCES

1. Liu Shih-yih, [Some problems of investigation of EEG and psychological mechanisms of brain physiology] (in Chinese), Acta Psychol. Sinica, No. 3, pp. 141-154 (1961).
2. Liu Shih-yih and Wu Qin-e, [Investigation of the EEG in 8- to 20-year-old subjects] (in Chinese), Acta Psychol. Sinica, No. 3, pp. 173-185 (1962).
3. Liu Shih-yih and Wu Qin-e, [Electroencephalographic and galvanic-skin investigation of the orienting reflex in man] (in Chinese), Acta Psychol. Sinica, No. 1, pp. 1-10 (1963). (English translation in this volume, pp. 62-74.)
4. Ivanov-Smolenskii, A. G., [On the interaction of the first and second signalling systems in some physiological and pathological conditions] (in Russian), Fiziol. Zh. SSSR, 35(5): 571-581 (1949).
5. Kozhevnikov, V. A., [Electroencephalographic study of the formation of temporary connections to sound stimuli in man] (in Russian), Candidate's Dissertation, Leningrad (1951).
6. Kratin, Yu. G., [The response of the alpha-rhythm to flashing as an indicator of the analyzing activity of the brain in man] (in Russian), Problems of Electrophysiology and Electroencephalography, pp. 49-59 (1960).
7. Liu Shih-yih, [New material on the question of temporary connection between stimuli of entero- and proprioreceptors] (in Russian), Scientific Reports of the Pavlov Institute of Physiology of the USSR, 2: 80-82 (1949).
8. Narbutovich, I. O., and Podkopaev, N. A., [The conditioned reflex as an association] (in Russian), Trud. Fiziol. Lab. I. P. Pavlova, 6(2): 5-25 (1936).
9. Peimer, I. A., [On the question of the physiological significance of the reaction of depression of alpha-waves of the electroencephalogram in man] (in Russian), Problems of Electrophysiology and Electroencephalography, pp. 70-79 (1960).
10. Podkopaev, N. A., [The conditioned reflex as an association] (in Russian), Material for the Fifth All-Union Meeting of Physiologists (1934), p. 62.

11. Bercel, N. A., Lowell, E. L., and Dossett, W. F., Conditioning of a photically driven cortical response, Electroenceph. clin. Neurophysiol., 11:613 (1959).
12. Gastaut, H., Jus, A., Morrell, F., Storm van Leeuwen, W., Bekkering, D., Kamp, A., and Werre, J., [Electroencephalographic characteristics of the formation of conditioned reflexes in man] (in Russian), Zh. Vissh. Nerv. Deiat., 7(1):25-38 (1957).
13. Gastaut, H., Roger, A., Dongier, S., and Regis, H., [Study of the electroencephalographic equivalence of the processes of central excitation and central inhibition during the elaboration of the conditioned reflex] (in Russian), Zh. vissh. Nerv. Deiat., 7(2):185-202 (1957).
14. Grastyân, E., Lissak, K., and Kekesi, F., Facilitation and inhibition of conditioned alimentary and defensive reflexes by stimulation of the hypothalamus and reticular formation, Acta Physiol. Hung., 9:133-150 (1956).
15. Jasper, H. H., and Shagass, C., Conditioning the occipital alpha rhythm in man, J. exp. Psychol., 28:373-388 (1941).
16. Jus, A., and Jus, C., Studies on photic driving conditioning in man, Electroenceph. clin. Neurophysiol., 11:178 (1959).
17. John, E. R., and Killam, K. F., Electrophysiological correlates of differential approach-avoidance conditioning in cats, J. Nerv. Ment. Dis., 131(3):183-201 (1960).
18. Knott, J. R., and Henry, C. E., The conditioning of the blocking of the alpha rhythm of the human electroencephalogram, J. exp. Psychol., 28:134-144 (1941).
19. Loomis, A., Harvey, E. N., and Hobart, G., Electrical potentials of the human brain, J. exp. Psychol., 19:249-279 (1936).
20. Magoun, H. W., Subcortical mechanisms for reinforcement, Electroenceph. clin. Neurophysiol., Suppl., 13:221-229 (1960).
21. Morrell, F., and Jasper, H. H., Electrographic studies of the formation of temporary connections in the brain, Electroenceph. clin. Neurophysiol., 8:201-215 (1956).
22. Morrell, F., and Ross, M. A., Central inhibition in conditioned reflexes, Arch. Neurol. Psychiat., 70:611-616 (1953).
23. Motokowa, K., Electroencephalograms of man in the generalization and differentiation of conditioned reflexes, Tohoku J. exp. Med., 50:225-234 (1949).
24. Sourek, K., and Skorpil, V., Use of galvanic skin reflex in neurosurgery, Electroenceph. clin. Neurophysiol., 13:141 (1961).
25. Stevens, J. R., Electroencephalographic studies of conditional cerebral response in epileptic subjects, Electroenceph. clin. Neurophysiol., 12:431-444 (1960).
26. Ulett, G. A., Stern, J. A., and Sines, J. O., Electrocortical changes during conditioning, Electroenceph. clin. Neurophysiol., 12:247 (1960).
27. Voronin, L. G., and Sokolov, E. N., Cortical mechanisms of the orienting reflex and its relation to the conditioned reflex, Electroenceph. clin. Neurophysiol., Suppl., 13:335-346 (1960).
28. Wang, G. H., Stein, P., and Brown, V. W., Changes in galvanic skin reflex after acute spinal transection in normal and decerebrate cats, J. Neurophysiol., 19:446-451 (1956).
29. Wells, C. E., The modification of alpha wave responsiveness to light by the juxtaposition of auditory stimuli, Electroenceph. clin. Neurophysiol., 12:238 (1960).
30. Yoshii, N., and Hockaday, W. J., Conditioning of frequency characteristic repetitive EEG response with intermittent photic stimulation, Electroenceph. clin. Neurophysiol., 10:487-502 (1958).
31. Yoshii, N., Matsumoto, J., Ogura, H. Shimokochi, M., Yamaguchi, Y., and Yamasaki, H., Conditioned reflex and electroencephalography, Electroenceph. clin. Neurophysiol., Suppl., 13:199-210 (1960).

APPENDIX A

Papers on Brain Research Published in the Acta Physiologica Sinica for the Period 1962-1966

1. Potentiation and depression effect of repetitive cortical stimulation on the evoked potential. Jiang Zhen-yu and Chang Te-hsing. Acta Physiol. Sinica, 25(1):29-35 (1962) [6].*
† 2. Cortical repetitive responses elicited by a single contralateral stimulus. Fan Shih-fang and Shen Ke-fei. Acta Physiol. Sinica, 25(2):114-118 (1962) [6].
3. Laminar distribution of gamma-aminobutyric and related amino acids. Chang Ke-pang, Chang Sheng-ken, and Chen Hsiu-fang. Acta Physiol. Sinica, 25(2):136-142 (1962) [6].
4. Studies on skin potentials and skin galvanic reflex in human subjects. Li Peng, Zheng Xia-zhao, Jin Wen-quan, and Zhao Xiu-ju. Acta Physiol. Sinica, 25(3):171-181 (1962) [5].
5. Topographic distribution of electrical responses of the olfactory bulb to olfacto-epithelial stimulations in the rabbit. Shen Ke-fei and Ho Shu-fang. Acta Physiol. Sinica, 25(4):244-249 (1962) [6].
6. Some observations on the changes of electrocortical activity in the process of formation of conditioned reflexes in the monkey. Wang Tai-an and N. I. Nezlina. Acta Physiol. Sinica, 25(4):291-298 (1962) [6].
7. Analysis of single unit activity in the lateral geniculate body of the cat. Liu Yü-min and Yang Chen-yü. Acta Physiol. Sinica, 26(1):12-21 (1963) [6].
† 8. Interaction of evoked cortical potentials in the rabbit. Zhang Gin-ru. Acta Physiol. Sinica, 26(2):165-171 (1963) [5].
† 9. Cortical responses to repetitive contralateral stimulation after sectioning of the corpus callosum. Fan Shih-fang and Shen Ke-fei. Acta Physiol. Sinica, 26(3):212-217 (1963) [6].
10. Distribution of glutamic acid decarboxylase, gamma-aminobutyrate-alpha-ketoglutarate transaminase and gamma-aminobutyric acid and their relationships in the rat brain. Chang Ke-pang and Chen Hsiu-fang. Acta Physiol. Sinica, 26(3):275-281 (1963) [6].
†11. Cortical excitability changes following transcallosal afferent excitation. Zhang Gin-ru. Acta Physiol. Sinica, 26(4):321-327 (1963) [5].
12. Observations on the ipsilateral evoked cortical potentials and their related afferent pathways. Liu Jin-long, Han Xian-wen, Xie Bi-xia, and Yang Hua-fang. Acta Physiol. Sinica, 26(4):379-386 (1963) [12].
13. A study of the components of the alpha-wave of human electroretinogram with special reference to the effect of light adaptation. Liu Yü-min and Yang Chen-yü. Acta Physiol. Sinica, 27(2):115-129 (1964) [6].
14. Two components of the scotopic beta-wave of the human electroretinogram. Liu Yü-min and Yang Chen-yü. Acta Physiol. Sinica, 27(2):130-139 (1964) [6].

---

*The number in brackets after each paper indicates the authors' institution in the list at the end of this appendix. In each instance, the author's surname appears first (see Introduction).
†Full translation appears in this volume.

15. Electroencephalographic changes during audiogenic seizures in albino rats. Shen Ke-fei and Ho Shu-fang. Acta Physiol. Sinica, 27(2):153-160 (1964) [6].
16. Primary electrical response to cessation of sound in the auditory zone of the cortex of the guinea pig. Lian Zhi-an. Acta Physiol. Sinica, 27(2):161-167 (1964) [6].
17. Spectral sensitivity of photopic alpha-wave of human electroretinogram. Liu Yu-min and Yang Chen-yu. Acta Physiol. Sinica, 27(3):219-225 (1964) [6].
†18. The interaction of callosal potentials and potentials evoked by thalamic stimulation. Zhang Gin-ru. Acta Physiol. Sinica, 27(4):348-355 (1964) [5].
19. Effect of chlorpromazine on electrical responses of cerebral cortex in rabbit. Chao Jung-jui, Hsie I-fang, Li Wen-chue, and Chang Shi-ru. Acta Physiol. Sinica, 28(2):109-115 (1965) [7].
20. Observations on the electroretinogram of the pure-cone eye of a species of tree squirrel. Liu Yü-min, Yang Chen-yü, and Li Jen-yuan. Acta Physiol. Sinica, 28(3):238-294 (1965) [6].
21. The pulse height analysis (amplitude spectrography) of the electromyography. Chen Wei-chang, Chu Chin, Chen Ta-kuang, and Chang Chi-cha. Acta Physiol. Sinica, 28(3): 286-294 (1965) [1].
22. The mechanisms of synchronization of skin potential wave with respiration. Li Peng, Cheng Jie-shi, and Sun Zong-han. Acta Physiol. Sinica, 28(4):378-386 (1965) [5].
23. The sleep and arousal effect of the anterior and posterior hypothalamus and their interaction. Lo Ai-lin and Yeh Chih-wen. Acta Physiol. Sinica, 29(1):5-12 (1966) [4].
†24. The effect of electric stimulation of the brain stem on the galvanic skin reflex. Li Peng, Cheng Jie-shi, and Sun Zhong-han. Acta Physiol. Sinica, 29(1):26-33 (1966) [5].
25. Effect of gamma-aminobutyric acid on the dominant excitation focus of the motor cortex created by anodal polarization. Wu Bao-hua and Wang Ching-wei. Acta Physiol. Sinica, 29(1):38-42 (1966) [6].
26. The effect of gamma-aminobutyric acid (GABA) on interoceptive conditioned reflex activity in different cortical regions of the dog. Mei Zhen-tong and Yu Hui-zhung. Acta Physiol. Sinica, 29(2):166-171 (1966) [6].
27. The effect of acetyl hydrazine on serum glutamic acid decarboxylase activity in the brain of the large white rat. Liu Gui-de and Wei Gui-hui. Acta Physiol. Sinica, 29(2): 219-222 (1966) [8].
28. Effect of noradrenalin and chlorpromazine on inhibition in conditioned reflexes in the large white rat. Xia Bing-nan, Zhou Wen-zheng, Song Jie-yun, and Cheng Cai-fen. Acta Physiol. Sinica, 29(2):223-228 (1966) [2].
29. Effect of acetylcholine and gamma-aminobutyric acid on the brain in audiogenic epilepsy in the large white rat. Chen Xiu-fang and Zhang Ke-pang. Acta Physiol. Sinica, 29(2): 229-233 (1966) [6].

Institutions in China from which These Papers Came

Huhehot

1. Inner Mongolia Medical College, Departments of Anatomy and Physics, and Inner Mongolia Hospital, Department of Physical Therapy (paper No. 21).

Kweiyang

2. Kweichow Medical Institute, Department of Pharmacology (paper No. 28).

Nanning

3. Kwangsi Medical College, The Central Laboratory (paper No. 12).

Peking

4. Institute of Experimental Medicine, Department of Physiology (paper No. 23).

Shanghai

5. First Medical College of Shanghai, Department of Physiology (papers Nos. 4, 8, 11, 18, 22, 24).

6. Institute of Physiology, Chinese Academy of Sciences (papers Nos. 1, 2, 3, 5, 6, 7, 9, 10, 13, 14, 15, 16, 17, 20, 25, 26, 29).

Taiyuan

7. Medical College of Shansi, Department of Physiology (paper No. 19).

Tsitsihar

8. Heilungkiang Medical Institute, Department of Pharmacology (paper No. 27).

APPENDIX B

Papers on Brain Research Published in English or in Russian in the Scientia Sinica for the Period 1952-1966

1. A contribution to the histology of mammalian cerebellum. Tsang Yü-ch'üan. Scientia Sinica, 1(1):95-101 (1952) [6].*
2. The striae medullares of the fourth ventricle as related to the nucleus arcuatus. Tsang Yü-ch'üan. Scientia Sinica, 2(4):301-311 (1953) [6].
3. The motor cell columns in the lumbosacral cord of sympodic fetuses. Tsang Yü-ch'üan. Scientia Sinica, 4(3):399-412 (1955) [7].
4. The motor cell columns in the cervical and lumbar enlargements of the spinal cord of the mole. Tsang Yü-ch'üan. Scientia Sinica, 4(4):583-590 (1955) [7].
5. Types and causation of the recurrent fibers in the mammalian cerebellar cortex. Tsang Yü-ch'üan. Scientia Sinica, 5(2):323-338 (1956) [7].
6. Special cell types in the mammalian cerebellum. Tsang Yü-ch'üan. Scientia Sinica, 5(3):535-542 (1956) [7].
7. Cortical response to antidromic stimulation of the corpus callosum. Feng Te-pei and Fan Shih-fang. Scientia Sinica, 6(1):159-168 (1957) [13].
8. Studies on proteins of the nervous system. I. A note on the electrophoretic behavior of some brain nucleoproteins. Sheng Pei-ken, Li Tsai-ping, and Tsao Tien-chin. Scientia Sinica, 6(2):309-316 (1957) [13].
9. The motor cell columns in the cervical and lumbar enlargements of the spinal cord of the mole-rat. Tsang Yü-ch'üan. Scientia Sinica, 6(2):327-338 (1957) [7].
10. Heterotopic Purkinje cells in the mammalian cerebellar cortex and their genetic significance. Tsang Yü-ch'üan. Scientia Sinica, 6(5):905-918 (1957) [7].
11. Isolation and properties of a new structural protein of muscle. Tsao, T. C., Hsü, K., Jen, M. H., Pan, C. H., Tan, P. H., Tao, T. C., Wen, H. Y., and Niu, C. I., Scientia Sinica, 7(6):637-647 (1958) [11].
12. On the nervous mechanism and changes of the acetylcholine content of the brain in experimental traumatic shock. Chen Chao-hsi. Scientia Sinica, 8(5):510-522 (1959) [2].
13. [Investigation of the conjoint activity of the cutaneous and visual analyzers of the rabbit] (in Russian). Mei Chen-tong and Chao Shang-chi. Scientia Sinica, 8(7):746-753 (1959) [12].
14. [Investigation of the conjoint activity of the cutaneous analyzer of the dog] (in Russian). Hsü Bing-hsüan and Mei Chen-tong. Scientia Sinica, 8(7):754-760 (1959) [12].
15. Electrical response of single neurons in the optic lobe of toad to photic stimulation. Chang Hsiang-tung, Chiang Chen-yü, and Wu Chien-ping. Scientia Sinica, 8(10):1131-1152 (1959) [12].
16. A decade of biology in China. Tung Ti-chow. Scientia Sinica, 8(12):1419-1444 (1959 [3].

*The number in brackets after each paper indicates the authors' institution in the list at the end of this appendix. In each instance, the author's surname appears first (see Introduction).

17. The vegetative functions of the corpus striatum. I. The change in blood pressure on stimulation of the caudate nucleus. Wang Pai-yang. Scientia Sinica, 9(3): 434-441 (1960) [10].
18. [Comparative permeability of the central nervous system to C^{14}-barbital and S^{35}-thiopental and mechanism of their action] (in Russian). Chu Shou-peng. Scientia Sinica, 9(7): 925-930 (1960) [1].
19. [The influence of the functional state of the cerebral cortex on the activity of gastric glands] (in Russian). Liu Tai-fong. Scientia Sinica, 10(5): 522-537 (1961) [9].
20. An analysis of fiber constitution of optic tract of cat. Chang Hsiang-tung and Cheng Tse-huei. Scientia Sinica, 10(5): 538-556 (1961) [12].
21. Interneuronal synapses in the human stellate ganglion. Tcheng Kuo-tchang. Scientia Sinica, 10(5): 557-568 (1961) [5].
22. The cerebellar connections of the mesencephalic nucleus of the trigeminus. Tsang Yü-ch'üan. Scientia Sinica, 10(7): 867-876 (1961) [6].
23. Neurons responsive to interruption of light and neurons active in darkness in the optic tectum of toad. Chang Hsiang-tung and Mkrtycheva, L., Scientia Sinica, 11(1): 90-99 (1962) [12].
24. [The effect of gamma-amino butyric acid (GABA) and chlorpromazine on conditioned-reflex activity of the rabbit] (in Russian). Mei Chen-tong and Chao Shang-chi. Scientia Sinica, 11(2): 241-250 (1962) [12].
25. Effect of monocular illumination on cortical response to optic nerve stimulation. Chang Hsiang-tung and Pressman, Y. M. Scientia Sinica, 11(9): 1249-1258 (1962) [12].
26. Neurotoxicity of streptomycin. Tsang Yü-ch'üan and Ch'in T'ing-ch'üan. Scientia Sinica, 12(7): 1019-1040 (1963) [6 and 8].
27. Apparent transformation from Felderstruktur to Fibrillenstruktur with stretch and the converse change with shortening in certain muscles of the chick. Yeh, Y., Huang, S. K., and Feng, T. P. Scientia Sinica, 12(8): 1242-1243 (1963) [12].
28. Changes of free amino acids content in stimulated nerve. Chang, S. K., and Feng, T. P. Scientia Sinica, 12(10): 1591-1593 (1963) [12].
29. Hypertrophy of "slow" muscle fibers following botulinum poisoning in chick. Feng, T.P., Wei, N. S., and Tian, W. H., Scientia Sinica, 12(11): 1757-1759 (1963) [12].
30. On components of α-wave of human electroretinogram. Liu Yu-min and Yang Chen-yü. Scientia Sinica 12(12): 1941-1943 (1963) [12].
31. A blue-sensitive component of scotopic b-wave of human electroretinogram. Liu Yu-min and Yang Chen-yü. Scientia Sinica, 12(12): 1943-1944 (1963) [12].
32. [The effect of cortical application of RNA on higher nervous activity] (in Russian). Mei Chen-tong. Scientia Sinica, 13(3): 524-525 (1964) [12].
33. Unit responses to sound stimulation and an inhibitory mechanism in nucleus corpus trapezoideum. Chang Hsiang-tung and Wu Chien-ping. Scientia Sinica, 13(6): 937-956 (1964) [12].
34. Stimulus-response compatibility and efficiency of information transmission. Hsu Lien-tsung, Yang Te-chuang, and Wang Tsi-chih. Scientia Sinica, 13(6): 1015-1017 (1964) [4].
35. Substitution of verbal reaction for motor reaction in signal identification tasks. Hsu Lien-tsung, Yang Te-chuang, and Wang Tsi-chih. Scientia Sinica, 14(1): 150-151 (1965) [4].
36. Selective reinnervation of a "slow" or "fast" muscle by its original motor supply during regeneration of mixed nerve. Feng, T. P., Wu, W. Y., and Yang, F. Y. Scientia Sinica, 14(11): 1717-1720 (1965) [12].
37. Local and stray-light components of the human electroretinogram due to stimulation by light subtending a 20° visual field. Liu Yu-min and Yang Chen-yü. Scientia Sinica, 15(5): 696-705 (1966) [12].

Institutions from Which These Papers Came

Lanchow

1. Lanchow Branch, Chinese Academy of Sciences (paper No. 18).

Peking

2. Chinese Academy of Medical Sciences, Department of Physiology (paper No. 12).
3. Chinese Academy of Sciences (Academia Sinica) (paper No. 16).
4. Institute of Psychology, Chinese Academy of Sciences (papers Nos. 34 and 35).
5. Institute of Zoology, Chinese Academy of Sciences (paper No. 21).
6. Peking Medical College, Department of Human Anatomy (papers Nos. 1, 2, 22, 26).
7. Peking Medical Institute, Department of Human Anatomy (papers Nos. 3, 4, 5, 6, 9, 10).
8. Peking Research Institute of Otology, Rhinology and Laryngology (paper No. 26).
9. Peking University, Department of Human and Animal Physiology (paper No. 19).

Shanghai

10. Fu-Tan University, Department of Biology (paper No. 17).
11. Institute of Biochemistry, Chinese Academy of Sciences (paper No. 11).
12. Institute of Physiology, Chinese Academy of Sciences (papers Nos. 13, 14, 15, 20, 23, 24, 25, 27, 28, 29, 30, 31, 32, 33, 36, 37).
13. Institute of Physiology and Biochemistry, Chinese Academy of Sciences (papers Nos. 7 and 8).

APPENDIX C

Some Linguistic Aspects of Scientific and Technical Chinese

Several systems based on the Roman alphabet (in the USSR, on the Cyrillic alphabet) have been devised for indicating the pronunciation of Chinese, the written form of which, employing ideograms (characters), is not self-pronouncing. In English, the most frequently encountered of these is the Wade — Giles system, which is usually used, for example, for important Chinese names, the dynasties, and for geographic names. The National Romanization system, devised in China in the late 1920's, has the advantage that the four tones (i.e., high-level, high-rising, low-dipping, and high-falling) of the Chinese spoken language, which otherwise would necessitate diacritical marks (¯, ´, ˇ, `) or superscript numerals for accurate pronunciation, are built into the system. The Pinyin (literally, "to spell") system of Romanization, which has been the standard in mainland China since 1957, represents a modified version of the National Romanization system, but without the complexity of the tonal spelling system of the latter. (The Pinyin system is employed in China as a complement to, rather than as a replacement for, Chinese characters.)

An example:

Characters	Wade — Giles	National Romanization	Pinyin	Russian
生理学报*	Shēng lǐ hsǘeh pào	Sheng lii shyue baw	Shēng lǐ xúe bào	Шэн лй сюе бào

Various Chinese dictionaries reflect not only different systems of romanization, but also different systems for ordering (and indexing) the characters themselves.

Renditions of Some Technical Terms into Chinese

Although many technical terms are rendered in Chinese as straightforward translations, others are rendered by transliteration, and still others by combinations of the two, in which indigenous Chinese terms may also be employed. The following are some examples (National Romanization System):

1. Midbrain reticular formation

   Mid brain net form structure
(Jong nao woang juanq jie gow)
中 脑 网 状 結 构

2. Medial geniculate body

   Internal knee body
(ney shi tii)
內 膝 体

3. Thalamic nucleus ventralis posterior

   Mound brain posterior belly nucleus
(chiou nao how fuh her)
丘 脑 后 腹 核

*Journal of Physiology (Acta Physiologica Sinica).

4. Hippocampus	Sea horse (hae maa) 海馬
5. Obex	Crossbar (shuan) 閂
6. Encéphale isolé	Isolated brain (gu lih nao) 孤立 脑
7. Absolute/relative refractory period	Absolute / relative no response period (jyue duey shiang duey bwu yinq chyi) 絕对 相对 不应 期
8. Spatial/temporal summation	Spatial / successive kind of summation (kong jian jih shyr shinq tzoong her) 空間 継时 性 总和
9. Evoked potential	Evoked electrical potential (yow fa diann wey) 誘发 电位
10. Orienting reflex	Orientation reflex (dinq shianq faan sheh) 定向 反射
11. Stereotaxic apparatus	Orientation instrument (dinq shianq yi) 定向 仪
12. Two-channel cathode-ray oscilloscope	Double channel cathode rays moving wave instrument (shuang shiann in jyi sheh shiann shyng bo chih) 双 綫 阴极 射 綫 示 波 器
13. Entropy	Entropy (fire quotient) (shang) 熵
14. Sodium pentobarbital	5th Chinese stem (transliteration) (wuh ba bii tuoo nah) 戊 巴比妥鈉
15. Chloralose	Chlorine aldehyde sugar (luh chyuan tarng) 氯醛 糖
16. Gamma-aminobutyric acid	Amino 4th Chinese stem acid (an ji ding suan) γ-氨基 丁 酸
17. Nikethamide	(Transliteration) (ni keh chah mii) 尼克刹米

An Example of Original Character Text, Romanization (National Romanization System), and Translation *

从丘脑記录反应的实験中，在观察結束以后，通过鋼針微电极通以阳极电流(約 1 毫安，0.5 分左右)。实験結束后，将脑子取出，固定在 10% 福尔馬林溶液中，数日后取出作厚片检查，按照 Sawyer 等人[2]的图譜找出电极尖端所在的位置，通电部分的脑組織一般呈深褐色。用这个方法測定的电极尖端位置和用亚鉄氰化鉀法[3]測得的相符。

实験在一电屏蔽保温的小室中进行，室温保持在 30℃ 左右。

Romanization

Tsorng chiounao jihluh faanyinq de shyryann jong, tzay guanchar jyesuh, tongguoh gangshen weidiannjyi tong yii yanjyi diannliou (iue 1 hauran, 0.5 fen, tzuooyow). Shyryann jyesuh how, jeang naotz cheuchu, guhdinq tzay 10% fueermaalin rongyeh jong, shuhryh how cheuchu tzuoh how piann jeanchar, jie jaw Sawyer et al. (2) de twupuu jao chu diannjyi jianduan de weyjyr. Tong diann buhfenn de nao tzuujy ibann cherng shern herseh. Yonq jeyg fangfaa tsehding de diannjyi jianduan weyjyr her yonq yah tiee chinghuahjiaa (3) tseh der de shiangfu.

Shyryann tzay idiann pyngbih bao uen de sheao shyh jong jinnshyng, shyh uen baochyr tzay 30°C tzuooyow.

Translation

Upon conclusion of the observations in the experiments in which responses were recorded from the thalamus, a direct current (of 1 ma, for approximately 0.5 min) was passed through the steel microelectrode, serving as the anode. After the experiment was concluded, the brain was removed, fixed in 10% formalin, and after several days, was removed, sectioned, and examined. The locations of the tips of the electrodes were determined with the aid of the atlas of Sawyer et al. (2), the parts of the brain tissue through which current had flowed usually showing a dark brown color. The locations of the tips of the electrodes measured by this method were in agreement with those determined by the potassium ferrocyanide method (3).

The experiments were carried out in a small electrically shielded and thermally insulated chamber, in which the temperature was maintained at approximately 30°C.

*From: Zhang Gin-ru, Interaction of evoked cortical potentials in the rabbit, Acta Physiol. Sinica, 26(2): 165 (1963) (Complete translation appears in this volume, on pp. 1–9.).

APPENDIX D

Dictionaries Used in the Preparation of These Translations

The linguistic complexities of Chinese (see Appendix C) give rise to a great variety of Chinese — English dictionaries and glossaries arranged in a number of different ways, and the existence of a number of Chinese — Russian and Russian — Chinese dictionaries, published both in Peking and in Moscow, widens the range of possibilities even more for translators who have a knowledge of the Russian language. Each translator undoubtedly has own preferences for dictionaries and glossaries, and the following compilation is simply intended as a listing of the ones in my own collection which have been used with greater (indicated by an asterisk) or lesser frequency in preparing the translations in this volume. Since English — Chinese (and Russian — Chinese) dictionaries are often helpful in verifying a meaning or in attempting to establish the exact shading, these are listed as well.

A number of these dictionaries (as well as many others) are included in the annotated listing of dictionaries and glossaries appearing in "Contemporary China — A Research Guide," by Peter Berton and Eugene Wu (The Hoover Institution, Stanford, 1967, 695 pp.); the number in brackets for each listing below indicates the corresponding entry in the above-mentioned reference work. Some others in the list were published too recently to have been cited in the above-mentioned guide. Two recently-published conversion tables for the various systems of romanization which include the Cyrillic (Russian) system are also included in the list (references 27 and 30).

When the need arose, standard Russian — English and English — Russian dictionaries (e.g., Smirnitskii, Levine, Muller, Rokitskii) were used, but special mention should be made of the Russian — English Dictionary of the Mathematical Sciences by Lohwater, published by the American Mathematical Society (1961).

<u>Chinese — English</u>

1. A Beginner's Chinese — English Dictionary of the National Language (Gwoyeu), 2nd rev. ed., W. Simon, London: Percy Lund, Humphries and Co., 1958, 880 pp.

* 2. Chinese — English Dictionary of Modern Communist Chinese Usage, 2nd ed. JPRS: 20,904, TT: 63-31674, U. S. Dept. of Commerce, Clearinghouse for Federal Scientific and Technical Information, Joint Publications Research Service, Washington, D. C., 1965, 845 pp. [440].

3. Concise Dictionary of Spoken Chinese, Yuen Ren Chao and Lien Sheng Yang. Harvard University Press, Cambridge, Mass., 1961, 292 pp.

4. Glossary of Physics, Science Press, Peking, 1964, 218 pp. [cf. 493].

5. Mathews' Chinese — English Dictionary, 3rd. ed. Harvard University Press, Cambridge, Mass., 1954, 1226 pp.

*Indicates dictionaries and glossaries most frequently used.

6. Mathews' Chinese — English Dictionary, Revised English Index. Harvard University Press, Cambridge, Mass., 1954, 186 pp.
* 7. Modern Chinese — English Technical and General Dictionary, 3 vol., McGraw-Hill Book Co., New York, 1963 (Supplement to Vol. III, 1965), 3864 pp. [481].

English — Chinese

8. Concise English — Chinese Dictionary Romanized, 3rd ed., James C. Quo, Charles E. Tuttle Co., Rutland, Vt., and Tokyo, 1962, 323 pp.
9. A Concise English — Chinese Dictionary, Commercial Press, Peking, 1965, 1252 pp.
10. Cousland's Medical Lexicon. Taipei, 1964, 588 pp.
*11. A Dictionary of Medical Terms (I-xue Ming-zu Hui-bian, or I-hsüeh ming-tz'u hui-pien). People's Hygiene Press, Peking, 1963, 761 pp. [541].
12. English — Chinese Dictionary of Mathematical Terms, Wang Chu-chi and others, The Commercial Press, Ltd., Hong Kong, 1964, 117 pp.
13. An English — Chinese Dictionary of Peking Colloquial, Sir Walter Hillier, enlarged by T. Backhouse and S. Barton, Routledge and Kegan Paul, London, 1953, 1030 pp.
*14. English — Chinese Medical Dictionary (New Medical Lexicon), China Publishing Co., Hong Kong, 1952, 1027 pp.
15. English — Chinese New Technical Dictionary of Electrical Engineering, Wan Li Book Co., Hong Kong, 1965.
16. English — Chinese Radio and TV Dictionary, Wan Li Book Co., Hong Kong, 1964, 417 pp.
17. International English — Chinese English through English Dictionary (based on Webster's New International Dictionary, 2nd ed.), Taipei, 1965, 1811 pp. [458].

Chinese — Russian

*18. Kitaisko — Russkii Slovar', I. M. Oshanin, ed., Institut Vostokevedeniia Akademii Nauk SSSR, Gosudarstvennoe Izdatel'stvo Inostrannykh i Natsional'nykh Slovarei, Moscow, 1959, 1100 pp. [445] (New edition in preparation.).
19. Kitaisko — Russkii Slovar': Nauchnykh i Tekhnicheskikh Terminov, ed. V. S. Kolokolov, Institut Nauchnoi Informatsii Akademii Nauk SSSR, Moscow, 1959, 568 pp. [486].
*20. Kratkii Kitaisko — Russkii Slovar', G. M. Grigor'ev and I. M. Oshanin. Gosudarstvennoe Izdatel'stvo Inostrannykh i Natsional'nykh Slovarei, Moscow, 1962, 631 pp. [447].

Russian — Chinese

*21. Bol'shoi Russko — Kitaiskii Slovar', Wang Tzu-yün and Liu Tse-jung, Commercial Press, Peking, 1961, 1384 pp. [460].
22. Kratkii Russko — Kitaiskii Slovar', Ya. B. Palei and V. K. Iustov, Gosudarstvennoe Izdatel'stvo Inostrannykh i Natsional'nykh Slovarei, Moscow, 1963, 590 pp. [462].
*23. Russko — Kitaiskii Politekhnicheskii Slovar', ed. Yu. M. Evsiukov and I. I. Moshentseva, Science Press, Peking, 1960, 1303 pp. [487].

Chinese — Chinese

*24. A Complete Index of Simplified Characters (Jianhuazi Zongbiao Jianzi), Language Reform Press, Peking, 1965, 64 pp.

25. Mandarin Dictionary (Dictionary of the National Language, or Gwoyeu Tsyrdean), 4 vols., Taipei, 1961, 4485 pp.

26. A Pinyin — Character Glossary (Hanyu Pinyin Cihui), enlarged ed., Language Reform Press, Peking, 1964, 669 pp. [cf. 429].

Other

27. Guide to Transliterated Chinese in the Modern Peking Dialect: Conversion Tables of the Currently Used International and European Systems with Comparative Tables of Initials and Finals, comp. and introd. I. L. Legeza, E. J. Brill, Leiden, 1968.
28. Russian — Chinese — English Chemical and Technical Dictionary, Scientific Information Consultants, London, 1965, 279 pp. [cf. 501].
29. Six Languages Dictionary, The Japan Times, Hara Shobo, Tokyo, 1958, 678 pp. [464].
30. Tables de Concordances pour l'Alphabet Phonétique Chinois, Centre de Linguistique Chinoise, École Pratique des Hautes Études, Mouton et Cie., La Haye, Paris, 1967.

APPENDIX E

CHINESE-LANGUAGE TITLES AND FOREIGN-LANGUAGE ABSTRACTS OF ARTICLES APPEARING IN THIS WORK

第26卷 第2期 生理学报 Vol. 26, No. 2
1963 年 6 月 ACTA PHYSIOLOGICA SINICA June 1963

家兎大脑皮层誘发电位的相互作用

張 鏡 如
(上海第一医学院生理学教研組)

INTERACTION OF EVOKED CORTICAL POTENTIALS IN THE RABBIT

ZHANG GIN-RU*
(Department of Physiology, First Medical College of Shanghai, Shanghai)

Abstract: It has been observed by Jarcho and by Chang that the cerebral cortex may undergo a postexcitatory depression and periodic excitability changes following an afferent excitation. The present work attempts to study further such excitability changes in the rabbit. The experiments were done on 32 rabbits anesthetized with chloralose and urethane. The interaction of the evoked cortical potentials was studied by means of paired stimuli applied at different intervals to different peripheral nerves. The results were as follows:

The cortical neurons in the focal projection area of a certain sensory nerve were found to undergo various degrees of postexcitatory depression following a conditioning shock applied to another sensory nerve which has an adjacent area of projection on the cerebral cortex. There were also periodic excitability changes accompanying the repetitive after-discharges of the conditioning evoked cortical potentials. Both the postexcitatory depression and the periodic excitability changes appeared more pronounced as the strength of the conditioning stimulus was increased.

The interaction appeared more marked when the potentials were led from the focal area of the conditioning response than from the fringes.

The postexcitatory depression was very weak when the conditioning shock was applied to the 12th intercostal nerve which has a negligible cortical projection.

When the conditioning stimulus was applied to the inferior alveolar nerve (consisting of mainly pain fibers) or the biceps femoris nerve (muscle stretch afferents), the results were similar to that obtained by applying the conditioning stimulus to a cutaneous mixed nerve, infraorbital nerve or sural nerve, for instance. The fact suggests that the excitability changes of the cerebral cortex following an afferent excitation fall into the same pattern regardless of the sensory modalities the stimulated nerve may subserve.

Homologous cortical areas of both hemispheres underwent similar excitability changes when the infraorbital nerve on either side was subjected to conditioning stimulation. Unilateral stimulation of the sural nerve could also affect the cortical excitability on both

* The author wishes to express his deep gratitude to Prof. H.-T. Chang, of the Institute of Physiology, Academia Sinica, for his very helpful guidance in the present work.

hemispheres, although the effect was much more marked on the contralateral than on the homolateral cortex.

Spatial summation of the postexcitatory depression has been observed when two nerves which have overlapping cortical projection areas were subjected to simultaneous conditioning stimulation.

Anesthesia tended to prolong the process of the postexcitatory depression and to abolish the periodic excitability changes.

The postexcitatory depression appeared more pronounced in Area II than in Area I of the sensory cortex.

第25卷 第2期 生理学报 Vol. 25, No. 2
1962 年 6 月 ACTA PHYSIOLOGICA SINICA June 1962

大脑皮层对于对侧皮层单个刺激的重复反应

范世藩　沈克飛
（中国科学院生理研究所，上海）

CORTICAL REPETITIVE RESPONSES ELICITED BY A SINGLE CONTRALATERAL STIMULUS

FAN SHIH-FANG AND SHEN KE-FEI
(*Institute of Physiology, Academia Sinica, Shanghai*)

Abstract: The cortical responses in unanesthetized rabbit to a single shock applied to the contralateral cerebral cortex show a train of surface negative waves at a frequency of 10—20 per second. Stimulation of the corpus callosum could also elicit such repetitive responses on both sides of the cerebral cortex; the responses were not synchronized. Transection of the corpus callosum did not abolish the repetitive responses, but additional transection of the thalamus along the midline abolished them. If one side of the thalamus was sucked off and the callosal connection kept intact, no matter which side of the cortex was stimulated, the repetitive responses could only be recorded from the side of the cortex with the thalamus intact. It appears that the generation of such repetitive responses is a result of the activation of some neuronic circuits involving the cerebral cortex and the thalamus by impulses coming through the corpus callosum or the bilateral connections of the thalamus.

第26卷 第3期 生理学报 Vol. 26, No. 3
1963年9月 ACTA PHYSIOLOGICA SINICA Sept. 1963

胼胝体切断后大脑皮层对于对侧皮层重复刺激的反应

范世藩 沈克飛
(中国科学院生理研究所，上海)

CORTICAL RESPONSES TO REPETITIVE CONTRALATERAL STIMULATION AFTER SECTIONING OF THE CORPUS CALLOSUM

FAN SHIH-FANG AND SHEN KE-FEI
(Institute of Physiology, Academia Sinica, Shanghai)

Abstract: A train of repetitive shocks applied to one side of the cerebral cortex in unanesthetized rabbit normally elicited a train of electrical responses of more or less regular amplitude in the contralateral cortex. After sectioning of the corpus callosum, the responses changed to a new one composed mainly of surface negative waves and characterized by periodic waxing and waning of the amplitude of the responses to individual shocks. Such responses could be obtained from the sensori-motor cortex as well as from the visual cortex. Further transection of the thalamus along the midline abolished such responses. Electrical responses could also be recorded from the central portion of the thalamus but with amplitude showing no waxing and waning phenomenon. The responses described above from the cortex and the thalamus could be degressed by stimulation of the divided corpus callosum on the recording side. It is assumed that stimulation of the cerebral cortex may activate the central portion of the thalamus which in turn activates the contralateral cortex. The nerve impulse passing through the corpus callosum may have a depressive action on such thalamically involved responses.

第26卷 第4期 生 理 学 报 Vol. 26, No. 4
1963 年 12 月 ACTA PHYSIOLOGICA SINICA Dec. 1963

刺激对侧大脑皮层对应点引起的皮层兴奋性变化

張 鏡 如
(上海第一医学院生理学教研組，上海)

CORTICAL EXCITABILITY CHANGES FOLLOWING TRANSCALLOSAL AFFERENT EXCITATION

ZHANG GIN-RU
(Department of Physiology, First Medical College of Shanghai, Shanghai)

Abstract: It has been observed by Chang and by Bremer that the cerebral cortex of the cat may undergo postexcitatory depression and facilitation following a transcallosal afferent excitation. The present work attempts to study further such excitability changes in the sensorimotor cortex of the rabbit, taking the cortical potentials evoked by stimulation of a peripheral nerve as the testing response. The results were as follows:

In the rabbit anesthetized with chloralose and urethane, the negative component of the testing response could be markedly facilitated by a shock applied to the homotopic point of the opposite cortex. The facilitatory effect lasted for about 1.2 sec., with its maximum occurring at about 150 msec. after the shock. The facilitatory effect on the positive component of the testing response was less marked and turned over quickly into a very prolonged secondary depression. There were also periodic excitability changes accompanying the repetitive after-discharges of the callosal response. In the 'encéphale isolé' preparation, the pattern of the excitability changes was similar to that mentioned above, but its time-course was much shorter.

The facilitatory effect required a suitable strength of the shock applied to the homotopic area of the opposite cortex. It was easily abolished by topical application of GABA, procaine or strychnine, but unaffected by barbiturate narcosis.

Laminar microelectrode analysis showed no particular layer which was concerned particularly with the facilitatory effect. However, laminar thermocoagulation of layers II and III did abolish the callosal facilitation.

Repetitive stimulation of the homotopic area of the opposite cortex resulted in severe depression which accompanied the shift of cortical D.C. potentials.

第27卷 第4期 生 理 学 报 Vol. 27, No. 4
1964年12月 ACTA PHYSIOLOGICA SINICA Dec. 1964

刺激丘脑体感觉核引起的皮层誘发电位和胼胝体电位的相互作用

張 鏡 如
(上海第一医学院生理学教研組,上海)

THE INTERACTION OF CALLOSAL POTENTIALS AND POTENTIALS EVOKED BY THALAMIC STIMULATION

ZHANG GIN-RU
(Department of Physiology, First Medical College of Shanghai, Shanghai)

Abstract: Previous workers have shown that the cortical potentials evoked by thalamic stimulation may be facilitated by preceding callosal potentials. This phenomenon is termed callosal facilitation by Bremer. The purpose of the present work is to find out whether the reverse is true, that is, whether callosal potentials can be facilitated by a preceding thalamic stimulation. Rabbits anesthetized with chloralose and urethane were used. The result shows that facilitatory effect can be observed on the slow component of callosal potentials. It lasts over 0.5 sec, with its maximum occuring at about 180 msec after the thalamic stimulation applied to n. ventralis posterior. It is more pronounced as the strength of the thalamic stimulation is increased. The facilitated callosal potentials are usually followed by very prominent repetitive after discharges and exert much stronger facilitating effect on subsequent cortical potentials evoked by thalamic stimulation. The callosal facilitatory effect is directly proportional to the amplitude of the slow component of callosal potentials. Tetanic stimulation experiments show that the first in the train of cortical responses to a short burst of repetitive shocks applied to n. ventralis posterior appears to be highest in amplitude if it is preceded by facilitated callosal potentials, whereas the third or fourth response is usually the highest in the absence of the preceding callosal stimulation.

第29卷 第1期
1966年3月
生理学报
ACTA PHYSIOLOGICA SINICA
Vol. 29, No. 1
March 1966

电刺激脑干对皮肤电活动的影响*

李鹏 程介士 孙忠汉
(上海第一医学院生理学教研組，上海)

心理学报
第 4 期
ACTA PSYCHOLOGICA SINICA
1965 年

不同年龄人脑 λ(Lambda)波的初步研究

蔡浩然 刘世熠

LAMBDA WAVES OF HUMAN SUBJECTS OF DIFFERENT AGE LEVELS

TSAI HAO-JAN, LIU SHIH-YIH

EEG records of Lambda (λ) waves of human subjects of different age levels indicated: 1) The appearance of λ-waves was most marked on Ss under 18 years of age, less obvious on subjects of 18 to 30 years of age, and least in Ss after age 31; 2) The λ-waves showed a higher frequency and a lower voltage and duration in adults than in children; 3) λ-waves more frequently appeared during visual perception, but not when the S engaged in visual imagination; 4) For subjects who manifested "on-response" and/or "driving response" in the occipital region, λ-waves were most easily recorded during visual perception.

*No foreign-language abstract was printed with this article.

心 理 学 报
第 1 期 ACTA PSYCHOLOGICA SINICA 1 9 6 3 年

人类定向反射的脑电与皮电图研究*

刘世熠 鄔勤娥

ELECTROENCEPHALOGRAPHIC AND GALVANIC-SKIN INVESTIGATION OF THE ORIENTING REFLEX IN MAN

LIU SHIH-YIH AND WU QIN-E

The present paper is devoted to the study, by means of the EEG and the GSR, of the "neural model" of the orienting reflex for various single brief stimuli (weak light, sound and touch). Our results showed the following:

(1) A brief single stimulus is an optimum afferent for study of the "model" of the orienting reflex. The "model" exhibits different features of response and of extinction for the different brief single stimuli.

(2) The appearances of the central (EEG) and efferent (GSR) components of the orienting reflex do not always concur with one another. The former is indistinct and the latter is prominent in children of less than 7 years age, whereas in adults the former is prominent and the latter is indistinct.

(3) It was shown that the orienting reflex can be both an unconditioned and a conditioned response.

心 理 学 报
第 1 期 ACTA PSYCHOLOGICA SINICA 1 9 6 3 年

人类暂时联系的脑电图研究*

刘世熠 鄔勤娥

ELECTROENCEPHALOGRAPHIC STUDY OF THE ELABORATION OF TEMPORARY CONNECTIONS IN MAN

LIU SHIH-YIH AND WU QIN-E

1. Evidence is adduced, by means of the EEG, of the possibility of elaboration of "conditionally-conditional" ("light + sound") temporary connections for pulsed single stimuli. The "conditionally conditional" paired stimuli were almost exactly simultaneous in their presentation.

* The foreign-language abstract for this article was originally printed in Russian.

2. The possibility is demonstrated of the elaboration of bidirectional "conditionally-conditional" temporary connections between the central (EEG) component and the efferent (GSR) component of the orienting reflex.

3. It was shown that conditioned reflexes of the "alpha-blocking" and "assimilation of rhythm" types could be formed in children and adolescent subjects who showed the phenomenon of "assimilation of rhythm."

INDEX*

* Brackets indicate the reference on the page indicated, if author's name does not appear.